GREEN PLANET

GREEN PLANET
How Plants Keep the Earth Alive

STANLEY A. RICE

RUTGERS UNIVERSITY PRESS
NEW BRUNSWICK, NEW JERSEY, AND LONDON

Library of Congress Cataloging-in-Publication Data

Rice, Stanley A., 1957-

Green planet : how plants keep the Earth alive / Stanley A. Rice.

p. cm.

Includes bibliographical references and index.

ISBN 978-0-8135-4453-3 (hardcover : alk. paper)

1. Plant ecology. 2. Vegetation and climate. 3. Plants, Useful. I. Title.

QK901.R53 2009

581.7—dc22

2008013964

A British Cataloging-in-Publication record for this book is available from the British Library.

Visit our Web site: http://rutgerspress.rutgers.edu

♻ This book is printed on recycled paper.

Manufactured in the United States of America

contents

illustrations

tables

acknowledgments

The author is grateful to Eldon Franz of Washington State University and Steward T. A. Pickett of the Institute for Ecosystem Studies for reviews and advice; to Gleny Beach of Southeastern Oklahoma State University for illustrations; to Sir Ghillean Tolmie Prance, former director of Kew Gardens, and Wes Jackson, founder of the Land Institute, for encouragement; to David Van Tassel of the Land Institute; Teresa Golden of Southeastern Oklahoma State University for help with photographs; to my agent Jodie Rhodes of La Jolla, California; and to editors Doreen Valentine and Beth Gianfagna, who made a tremendous contribution to producing a readable and interesting manuscript.

GREEN PLANET

INTRODUCTION

REMNANTS OF PARADISE

The alder, whose fat shadow nourisheth
Each plant set neere to him long flourisheth.
—WILLIAM BROWNE, CA. 1613

It certainly didn't seem like paradise. I was up to my hips in the sucking slime of a swamp. Because of the water had very little oxygen, the leaf litter and the corpses of mosquitoes and snapping turtles did not completely decompose. Instead they produced a dark brown glue that stained my clothes like a stygian tea. Bacteria released a putrid scent of hydrogen sulfide. I had no idea how deep the muck was, and my left leg slipped in more deeply as I tried to lift my right. There was nothing to grab on to except poison ivy, the thorny branches of greenbrier and rose, or dead sticks.

I was in Hudson Pond in central Delaware. The bridge for U.S. Highway 113, only a few yards from where I stood, rumbled with hundreds of cars and trucks. The passing motorists who looked down on me might have thought I was crazy if they knew that I was studying the small trees that grew in the swamp. If they knew that I had driven a thousand miles to see them, their suspicions would have been confirmed. But when I looked up at the trees that I had driven so great a distance to see, it all seemed worthwhile. There they were, *Alnus maritima*, the seaside alders. Each one consisted of a cluster of little gray trunks, with serrated leaves and puffy, conelike fruits (fig. I.1). What makes this species so special is that it is very rare, and that it has an unusual geographic distribution. All of the seaside alders in the entire world occur in just three populations. The first population lives along

FIGURE I.1. Seaside alders, such as this tree in the middle of the Blue River in Oklahoma, are nearly extinct. Many plant species like this one are facing extinction before their importance in their habitats or to the human economy have been adequately studied. Photograph by the author.

clear, rushing streams in two counties in Oklahoma. The second inhabits a single swamp in northwestern Georgia. The third is found on the Delmarva Peninsula east of Chesapeake Bay. I had visited the first two groups and was now exploring the third. Each of these populations is separated from the others by about a thousand miles.

Part of the mystery of the seaside alder is how this species came to exist in only these three places. It apparently had a wider range in the past and then died out everywhere except in these locations. Another species of alder, the hazel alder (*Alnus serrulata*), lives right alongside the seaside alder in Georgia and Delmarva—in fact, while standing under a seaside alder, I could look a few yards to one side and see a hazel alder, in exactly the same habitat conditions. But there are millions of hazel alders throughout the eastern United States. Why was the seaside alder dying away, while the hazel alder was thriving? My colleagues and I have determined that it was because hazel alders can survive in the shade, and seaside alders cannot—and most swamps are shady.[1] Another aspect of the puzzle is that the alder species that are most genetically similar to the seaside alder are not other American alders, but species of alder that live in Asia, such as *Alnus nitida*. The story of the seaside alder is a complex and fascinating story for me and my fellow botanists to investigate.

Certainly botanists find the seaside alder interesting, but why would anyone else care about it? Why should *you* care about it? Why should you care about any other species of grass, wildflower, bush, or tree? These are reasonable questions. The purpose of this book is to answer them.

A basic knowledge of what plants are and what they do is essential for everyone on the earth—not just for botanists, gardeners, farmers, and landscapers. Plants make life on Earth possible. All the oxygen in the air has come from plants. Plants help to prevent an excessive greenhouse effect by removing carbon dioxide gas from the air. Plants put moisture into the air and create cool shade, in which animals like us can survive. Plants hold down the soil and allow water to penetrate into the ground, thereby preventing floods and mudslides. All the food in the world, for all animals—including humans, is made by plants. Plants create the habitats in which all species live, and they allow these habitats to recover from disturbances. Plants are also the essential basis of the human economy. Wetlands, such as the swamp in which I stood, are not simply places that are wet. They help to prevent flooding and soil erosion, and it is the plants that live in them that provide these essential life-support services.

Plants not only keep this planet alive but have, over the course of millions of years, made Earth what it is today. As climatologist David Beerling says, plants are not "silent witnesses to the passage of time." Plants, directly and indirectly, have altered the amount of oxygen and carbon dioxide in the atmosphere for the past 400 million years.[2] However, if the seaside alders disappeared, there are plenty of other plants that can make food and oxygen, and plenty of other wetland plants that can prevent floods and water pollution. North America has six other species of alder. This being the case, why should anyone care about this particular species?

There is always the possibility that a plant species, such as the seaside alder, may promote human health or provide other economic benefits. Investigations in our laboratory have shown that the seaside alder appears to have medicinally active bark. Maybe, after many more years of research, we will be able to use the seaside alder as the basis for new treatments for diseases. And maybe not. There is also the possibility that the seaside alder may play a crucial role in the wetlands that it inhabits, a role that no other species of plant could quite fill. And maybe not.

This is perhaps the answer to the question of why we should save the seaside alder: we don't know what it might be worth. Every time we allow a species to slip into extinction, we are throwing away a biological component of the world that might or might not be of crucial importance. Just because we do not know whether a species is useful to the human economy or to the community of species of which it is a part does not mean that we can cast it aside.

In this book, I first examine how human activities have greatly altered the face of the earth, particularly the forests and fields that cover it. Next, I present a ten-point argument that plants do not just cover the earth but keep it alive; I devote a chapter to each of these points:

1. Plants have produced nearly all of the oxygen that is in the atmosphere.
2. Plants remove carbon dioxide from the atmosphere, helping to regulate the greenhouse effect.

3. Plants create cool shade.
4. Plants help to prevent floods and droughts.
5. Plants put sunlight energy into the entire food chain of the earth.
6. Plants transform dirt into soil.
7. Plants create the habitats in which all organisms live.
8. Plants grow back after disturbances, restoring habitats.
9. Plants are the basis of agriculture.
10. The diversity of plant species sustains natural habitats and human activities.

Economist Robert Costanza and colleagues have estimated that the "ecosystem services" provided by the natural world, mostly by plants, and for which we do not have to pay, are worth $33 trillion to the world economy—an amount equivalent to about half of its total annual productivity.[3] Then, in the final chapter, I present a vision of what we can do to enable plants to help rescue us from the environmental disasters we have brought on the earth.

During our scientific studies, my colleagues and I have found that a single seaside alder single tree may produce thousands of seeds each year and that these seeds produce healthy young plants in the greenhouse. Yet we have seldom observed an alder seedling in the wild. Old trunks die, and new ones grow back from the rootstock, but the alders do not seem to spread. The surviving seaside alder clusters may therefore be hundreds or thousands of years old, and the death of these clusters would mean the extinction of the species. Humans could destroy the entire species by damaging the wetlands in which they live. We could drain and fill in their wetland habitats. Housing construction has destroyed wetlands in Georgia where the seaside alders used to live. Or we could poison them with air pollution. While standing in Hudson Pond, I observed many alder leaves that were damaged by vehicle fumes from the nearby highway. Or we could destroy them by changing the climate and water resources on which they depend. In Oklahoma, global warming and depletion of underground water resources may cause their current habitat to become too hot and dry for their survival. These sentinels from an ancient world could vanish quickly.

Botanists like myself are on a mission: to proclaim to the world the importance of plants and why we should save those that are endangered. We need to preserve the wild plants of forests and fields as if our lives depended on it—because they do. The greatest extinction of plant species that has ever occurred is now under way, and it is unnoticed by most people. The remnants of paradise are not just in distant rain forests or islands, but along fishing creeks in Oklahoma, in rural swamps in Georgia, and beside a highway in Delaware. The lone man standing out in the swamp muck is trying to, in the small way that any one person can, save the world that trees and bushes and grasses and wildflowers keep alive.

chapter one

AN INJURED PARADISE

I want to tell what the forests
were like
I will have to speak
in a forgotten language
—W. S. MERWIN

About seven thousand years ago, much of the earth was a paradise. Many cultures have legends of a primordial paradise. One of these legends, familiar to people of the Western cultural tradition, is the Garden of Eden. This story provides an image of the original relationship between humans and the natural world: Eden was a garden planted by God himself, in which Adam and Eve, the archetypes of humankind, lived. It was a garden of inexpressible beauty in which even God enjoyed taking a stroll in the cool of the morning. A mixture of forested orchard and herbaceous field, Eden contained every tree and plant that was good for food and pleasing to the eye. That's a fair picture of what the earth was like just before civilization.

The world had not always been like this. At the end of the most recent ice age, about twelve thousand years ago, the northern continents experienced violent conditions. Glaciers melted and retreated northward, exposing land that had been scraped bare of life and topsoil. Strong winds carried dirt in massive dust storms. The glacial meltwater produced huge rivers. For example, the part of the Missouri River that today is less than a mile wide was ten miles across when the glaciers were retreating. In some cases, ice fragments blocked the water into vast lakes. When the ice blocks melted a little, floods of cold water would erupt from the lakes and scour thousands of square miles of land downstream.

But by about seven thousand years ago, warmth had returned; indeed, much of the earth was warmer than it is today. The forests and grasslands returned.[1] The forests had assumed roughly their present geographical locations long before human civilization. For our species, these forests and fields were a type of Eden. They could not be considered a zoo, for many species of large mammals (most famously the mammoths and mastodons) became extinct right at the end of the ice age, but the forests and fields were an abundantly beautiful and productive garden. From this time right up to the present day, the climate of the earth has not undergone the sudden and violent changes that were so common during the past two million years. Archaeologist Brian Fagan has called these millennia of nice weather "the long summer."[2]

The extent, and the appearance, of the forests just before civilization would have astonished us modern people if we could have seen them. Most noticeable would have been the large trees, many of them with trunks a yard or more in diameter. Today, when people visit remnants of ancient forests, they can be overwhelmed by these large trees.[3] Examples include Tane Mahuta, the largest of the kauri trees of New Zealand (*Agathis australis*), whose trunk, at 150 feet above the ground, is 15 feet thick; the General Sherman tree in California, the largest giant sequoia (*Sequoiadendron giganteum*), whose trunk is 36 feet thick at the base and still 14 feet thick at a height of 180 feet (fig. 1.1); one of the twenty-six California coast redwoods (*Sequoia sempervirens*) known to be over 360 feet high; the giant baobabs at Morondava, Madagascar; the angel live oak (*Quercus virginiana*), with a massive crown of branches reaching to the ground on John's Island, South Carolina; or some of the giant trees of Japan, such as the Jomon Sugi of Yakushima, the largest and oldest cedar (*Cryptomeria japonica*) in Japan. But trees of this stature were relatively common in the primordial forests that blanketed the earth after the most recent ice age. The forests also had healthy crops of small trees, because small fires created patches of open soil in which the trees and many other plant species could regenerate. Now, however, even these few remaining large trees, which have escaped the ax and are protected by national governments, may not be able to survive the higher temperatures and drought associated with global warming (see chapter 3). A couple of generations from now, they may be only a memory.

These primordial forests were found throughout much of the world. Much of England, which today consists mostly of open fields, was a Sherwood Forest; Strabo, a Roman chronicler, described Britain as "overgrown with forests." Italy, today a land of grass and shrubs, had deep, dark forests. The Antium and Avernian woods were to the south, and the Ciminian woods to the north, of Rome. All of these woods were described as impenetrable and filled with hostile tribes, and commercial travel through them was forbidden in early Roman times. Northern Africa had extensive forests and was renowned as a source of lumber during the days of the Roman Empire. The hills of what are now Syria, Turkey, Iraq, and Iran had extensive coniferous forests. Woodlands covered the now arid landscapes of Greece, Crete, and Cyprus.[4] Lebanon was the home of vast groves of cedars, biblical symbols of strength and grandeur. Extensive forests covered China. The great swaths of tropical rain forest formed a green sash around the earth's equator, across South America, Africa, southeast Asia, and northern Australia. India was a land of jungles.

All over the world, there were abundant resources for a life of hunting and gathering. Many of these resources were in the forests, whereas others were in grasslands. Gathering tribes found bountiful supplies of wild wheat and barley in the ancient Middle East. Tribal peoples knew how to find and prepare hundreds of kinds of fruits, roots, nuts, and grains. Although the hunter-gatherer lifestyle is certainly not a paradise—warfare seems to be a constant condition in these societies—their environment was nearly a paradise. The forests and grasslands were the ecological niche of our species. In many ways, our bodies and minds are still adapted to that world. We have changed it drastically, and we miss it.

Tribal humans had a significant impact on the natural forests and fields. Mass extinctions of large mammals occurred about fifty thousand years ago in Australia, about twelve thousand years ago in North America, and about two thousand years ago in New Zealand and Madagascar—precisely at the times when the first humans arrived in these places.[5] Once this happened, people adopted hunting practices

FIGURE 1.1A. The General Sherman Tree is the largest of the giant sequoias of the Sierra Nevada in California. Many forests around the world used to contain very large and ancient trees (though seldom as large as the sequoias). Most of these forests have been cut down. Although many have grown back, and most of their tree species have returned, the woodlands consist largely of younger and smaller trees. Photograph by the author.

that prevented the exhaustion of the prey populations. Hunting by humans may have been the single most important factor in the control of the prey populations. Scientists have recently begun to realize that the prodigious herds of bison and flocks of passenger pigeons in

FIGURE 1.1B. Continued

eighteenth-century North America were population explosions that occurred when European diseases began to decimate the Native American hunters that had previously kept them in check.[6] When botanist William Bartram traveled in Florida in the late eighteenth century, he observed many hundreds of alligators at a time, perhaps because of an explosion in fish populations.[7]

Hunters and gatherers often altered and managed their habitats. Some of these alterations were merely extensions of natural processes. For example, Native Americans used fire as a management tool. When English settlers arrived at Jamestown in 1607, they found that Native Americans used fire to clear out forest undergrowth, with the result that a man on horseback could easily ride through these forests. Other Native Americans started fires in grasslands, which, like the natural fires, would destroy trees and encourage grasses to resprout. It has even been suggested that natives of eastern North America, who liked chestnuts, would plant the nuts to ensure a continued supply—and that this may have been an important reason that the American chestnut (*Castanea americana*) dominated such an extensive area of forest.[8] But the impacts of hunting, fire, and planting did not diminish the range of the forests or the magnificence of its largest trees.

Then, all around the world, humans began to practice agriculture. Agriculture was preceded by a long period of coevolution between human cultures and the wild plants that they preferred.[9] But some environmental trigger caused humans to begin planting, rather than just managing, their preferred wild food plants. In the Middle East, it may have been when the world, warming from a recent ice age, experienced one of its temporary reversals back into cold, dry conditions. Agriculture began first in regions, such as the Middle East, in which the wild food plants such as wild wheat and barley had characteristics suitable for agricultural manipulation. Because of the absence of easily domesticated wild plants, it took longer for the natives of Mexico and South America to bring maize and potatoes into cultivation and for the natives of China to develop rice paddies. Herding also began about the same time; it started earlier in the Middle East, where many species of wild mammals (such as sheep, goats, and cows) were behaviorally suited to being herded, than in Central and South America.[10]

As will be explored in chapter 10, agriculture changed the world. Even though it reduced the diversity of the human food base and resulted in poorer nutrition for the average person, agriculture did allow a vast increase in the amount of food that could be produced, thus allowing the human population to grow and civilization to begin.

The biblical story of Eden describes this transition in the Middle East as a "fall." Adam and Eve, driven out of the garden, had to survive by tilling the ground. Agriculture is hard work—the biblical story calls it "the sweat of your brow"—compared with which hunting and gathering is relatively easy work.

In the biblical story, the garden was protected by a flaming sword and cherubim so that Adam and his descendents would stay out. What this may symbolize is that once civilization had begun, there was no going back. Agricultural human populations were too large to ever again revert to hunting and gathering. Once human populations had grown, farmland had become valuable, and armies defended these societies, a return to the days of hunting and gathering was no longer possible. The only way a population could revert to hunting and gathering was after a catastrophic collapse of its agricultural civilization.

Agriculture spread from several points of origin until it penetrated nearly the entire habitable world. Middle Eastern agriculture spread eastward as far as northern India, and northward into Europe, which did not develop its own native agriculture. Rice cultivation spread from China into southeast Asia and Japan. As agriculture spread, people destroyed many of the forests and changed the landscapes into the appearance that they have today.[11]

The spread of agriculture did not always result in the destruction of the forests. In North America, the native tribes practiced extensive agriculture. Maize, beans, and squash from Mexico spread northward into North America, where they replaced the sump weed and sunflowers that the natives had previously grown. The early Spanish explorers in what is now the southeastern United States and the English settlers in Jamestown encountered tribes that raised large fields of corn. The forests were destroyed when the fields were cleared, but the natives abandoned the fields after a few years and allowed secondary forests to grow back (see chapter 9).[12] After a few decades, they would clear the secondary forests (a task much easier than clearing primary forests using stone tools) and plant them again. As a result of these practices, many primordial forests with large trees, especially on mountain slopes, remained intact, and secondary forests were abundant in North America.

When botanist William Bartram traveled through southeastern North America in the 1770s, he visited many Seminole, Muskogee, and Cherokee villages that practiced extensive agriculture—but mostly on the floodplains, leaving the mountain slopes intact.[13] Nor did the spread of agriculture require extensive degradation of the soil. Terraces, used to raise rice, trapped soil as well as water, and allowed continuous cultivation in extensive areas of China, southeast Asia, and Japan.

There are even some instances in which the practice of agriculture built up rather than depleted the soil. The soil of the Amazon rain forest is notoriously poor, except along rivers and near volcanic slopes where minerals can be replenished. Even today, people cut down forests in the Amazon and try to raise crops, a practice that is only marginally and temporarily successful. However, when Francisco de Orellana sailed up the Amazon in 1542, he noted that there were extensive towns, supported by agricultural fields. Long considered fictitious, Orellana's accounts have now been given historical credibility by the discovery of black soils (*terra preta de Indio*) in the Amazon, which are filled with shards of pottery that were used by natives during the centuries preceding Orellana's expedition. Apparently the natives had constructed mounds of potsherds and compost to produce an artificial and rich soil.[14] The natives of what is now the southeastern United States apparently did the same thing. During his travels, William Bartram frequently encountered ancient mounds of broken pottery and animal bones, which produced rich farmland above the floodplains. The Muskogee and Cherokee tribes whom he visited had conquered earlier tribes; neither they nor the remnants of the earlier tribes knew who had built the mounds.[15] The mounds may have been built during the Cahokian or Mississippian culture about 1000 CE, appropriately called the period of the Mound Builders. They are most famous for the ceremonial mounds in their villages, but agricultural mounds were even more important in some regions. The Hohokam civilization, which flourished about 1000 CE in what is now Tucson, Arizona, transformed their landscape with irrigation check dams, as well as terraces and rock piles for planting crops.[16]

But usually the spread of agriculture and civilization was destructive to the fields and forests. All around the world, cities grew, empires spread, wild fields became cropland, and the forests became lumber. This practice was only temporarily successful. As we see in the following examples, civilizations often collapsed at the same time as a naturally occurring drought that caused the interruption of agriculture.[17] Had agricultural practices not destroyed the fields and forests, these civilizations might have been able to survive the droughts that ended up destroying them. Then the landscape would recover—after the collapse of the civilization.[18]

Ancient Sumeria was perhaps the world's earliest empire. After a period of prosperity in the area surrounding the city of Akkad, the empire that had been consolidated under King Sargon disintegrated, and many of its cities were abandoned. A poem, *The Curse of Akkad*, was written soon after the fall of the empire. It ascribes the collapse of Akkad to Enlil, the god of wind and storms, who took revenge against one of Sargon's grandsons for disrespecting one of his temples. The poem said that the fields produced no grain, the orchards no fruit, and the rivers no fish, resulting in mass starvation. Traditionally considered fiction, *The Curse of Akkad* may have a basis in fact. Scientists have found layers of eroded soil in the archaeological record, and terrestrial sediments in the Persian Gulf, dating to just after the end of the Sumerian Empire. Apparently a drought, which occurred over a large region, was the primary cause of agricultural failure. The destruction of forests and loss of soil from the fields left the Sumerian agricultural economy fatally vulnerable to such a natural disaster.[19] Some of the changes in the agricultural economy, such as the shift from wheat to the more drought-tolerant barley, may have been primarily a response to the drought. However, during that time, the use of wood for charcoal decreased, while the use of dung for fires increased. Spiny and unpalatable plants then replaced good forage plants in the pastures of the ancient Near East. This is clear archaeological evidence that humans induced the degradation of the land by depleting the wood from the forests and overgrazing the fields.[20]

Deforestation spread through the Mediterranean region. The Romans cut down the forests of Italy, while conquering the tribes that lived in them. The sandarac, a type of pine that dominated the forests of northern Africa, was highly prized by the Romans, who cut them all down by the middle of the first century CE. The forests of Greece, Crete, and Cyprus fell to the ax, for agriculture and for smelting metals.[21] Even the legendary cedar forests of Lebanon were destroyed, many of them (according to the biblical stories) to build Solomon's temple, many others for building the Phoenician fleets. Today, the Mediterranean basin, particularly the Near East and northern Africa, has few forests. There is also evidence for environmental degradation in Europe during the decline of the Roman Empire.[22]

Meanwhile, agricultural civilizations experienced disasters in North and South America as well. Mayan civilization in Central America disintegrated during a period of war between cities. These wars may have resulted from a shortage of resources. A layer of lake sediments indicates that a prolonged drought occurred at that time. But even before the drought, there was deforestation and erosion; much farmland was abandoned, and indicators of health, still visible in skeletons from that time, showed a phase of malnutrition that preceded the drought.[23] South American civilizations also fell long before the arrival of the Europeans—the Moche of the Peruvian coast, and the Tiahuanaco of the Andes—apparently also as a result of the loss of natural forest and soil resources. The Cahokian civilization, centered in what is now Missouri, collapsed during an era of flooding that may have been caused by deforestation. The basin of what is now Tucson was once the irrigated farmland of the Hohokam civilization, which also collapsed.[24]

The Chaco Canyon archaeological site is today isolated in the desert of northwestern New Mexico. During my visit to this site, I walked in the blistering dry heat among the walls of Pueblo Bonito, reconstructed from original stones. I saw the remains of what had once been a flourishing civilization. Many adjoining apartments and numerous ceremonial centers (including circular kivas) had once housed hundreds of people (fig. 1.2). These people had required a large resource base of water, food, and wood. The water that they drank came from the nearby stream, as did the irrigation water for the maize that they ate. They needed wood to reinforce the buildings. Piñon pine wood had originally supported the walls and roofs. The roofs have completely vanished, but the roof beams had been

FIGURE 1.2. Piñon pine forests once surrounded the civilization now known as Chaco Canyon in New Mexico. Today, a drier climate prevents trees from growing here. Drought caused the collapse of this civilization, but deforestation and erosion from logging and agriculture contributed to the impact that the drought had on a culture that might otherwise have persisted. Photograph by the author.

supported by holes in the nearby cliffs, still visible today. Therefore, piñon woodland and abundant water had once been present. Today there are no piñons anywhere close to Pueblo Bonito. I saw only desert bushes and a nearly dry stream that supported just a couple of cottonwood trees. Once again, a change of climate was the primary cause, but the people contributed to the end of their own civilization by cutting down most or all of the trees.[25] Archaeologists estimate that more than two hundred thousand trees were necessary to build the Chacoan civilization. Just before the collapse of their civilization, the builders had run out of piñon wood and had to import ponderosa pine from far away in a last, desperate effort to maintain their way of life. Pollen deposits indicate that the pollen of desert shrubs replaced that of conifers prior to the collapse of this civilization. A similar catastrophe occurred at the site of what is now Mesa Verde, in southwestern Colorado. When the ancestors of the modern Navaho and Apache arrived, the cities of the previous inhabitants (whom they called the Anasazi) were already abandoned. These collapses were not completely the result of a generalized drought, because they occurred at different times in different places: the establishment of the civilization in what is now Bandelier National Monument occurred about the time that the nearby Mesa Verde civilization was declining, and then it, too, collapsed.[26]

In some places, as populations grew and people cut down the forests, government officials could see what was happening and took moderately successful steps to protect the woods—not an intact forest for the sake of nature, but as a resource for continued exploitation. Two examples are medieval Europe and seventeenth-century Japan.

As European populations grew during the thirteenth century, deforestation accelerated. The nobility set aside forests for various purposes. In some cases, it was so that they would have private hunting grounds for their own amusement. In other cases, it was to assure the continued availability of large trees for shipbuilding and construction. They restricted peasants from cutting old forests that contained large trees and from cutting new forests that had not yet grown enough. Large oak trees that produced seeds for forest regeneration were protected by law. Because saws could cut larger trees more quickly and quietly than hatchets, in some areas the use of saws was prohibited. Growing populations and restrictions on woodcutting resulted in extensive temptations to conduct illegal forest activities such as lumbering and the making of charcoal. The position of forester for a nobleman was itself a life of moderate luxury; the forester hired sheriffs who enforced the laws governing the use of the forest. These measures were only partially successful. In the fourteenth century, the Black Death (a massive epidemic of bubonic plague) killed about one-third of all Europeans and caused massive cultural and economic disruptions. It was the Black Death that allowed the forests to regenerate.[27]

In Japan, about 1670, the newly centralized Shogunate government undertook a vigorous forest conservation program, including an inventory of nearly every tree in some of the forests, mainly to ensure that wood remained available for building. Scholars wrote sylvicultural manuals, and forestry "missionaries" traveled village to village to spread the news of how to save the forests. Foresters established plantations of sugi and hinoki cedars. This was not the first time that trees had been planted and preserved in Japan. A forest near Kumano is actually a plantation, but looks like a virgin forest because it has not been cut since 1391. But beginning with the Meiji Period, forest management and preservation became national policy, and deforestation was largely controlled.[28]

When Europeans began to spread throughout the world, they found forests that they could quickly exploit. When they first came to North America in the early sixteenth century, they established a few struggling outposts and depended on Native Americans for their survival. But the Europeans also brought diseases, which spread from one native village to another and partially or completely depopulated them.[29] After the collapse of these populations (perhaps as many as 90 percent of them died), the Native Americans could no longer maintain the forests by the controlled use of fire. Small trees grew back underneath the large ones. When many more Europeans arrived, in the seventeenth century, they found woodlands thick with undergrowth, and assumed that the forest had always been this way (what the American poet Henry Wadsworth Longfellow called "the forest primeval").

One of the first things the Europeans did upon reaching North America was to cut down as many trees as they could, not only for lumber and to clear farmland, but to destroy the habitat of the remaining natives. Unlike the Native Americans, the Europeans and then their American descendants practiced continuous agriculture, so the forests did not grow back; agriculture spread only by leveling more and more forests. During the nineteenth and early twentieth centuries, almost all of the North American deciduous forest, and much of the coniferous forest, fell to the ax and saw. Some forests, such as the Cross Timbers of Oklahoma and Texas, escaped destruction, largely because their lumber was inferior for construction and their soils unsuitable for agriculture.[30]

Meanwhile, well into the twentieth century, there was very little destruction of the tropical forests. These forests, too, had experienced fluctuations of fortune during the ice ages. Although neither glaciers nor cold weather reached into the tropics, the ice ages (when sea levels were three hundred feet lower than they are today) were a time of relative drought. Many tropical forests had retreated into small wooded patches surrounded by grassland. When the rains returned a few thousand years ago, the forests spread back over the equator. The tropical forests, like the temperate ones, had survived and recovered from massive disruptions.

As late as the 1960s, almost all of the tropical forests remained intact. They were heavily populated by tribal peoples who not only hunted and gathered, but also carried out shifting agriculture, which was more like gardening than agriculture and had little permanent impact on the forest.[31] Europeans and Americans had established "banana republics" and rubber and coffee plantations, but these affected relatively small areas. Starting in the 1960s, the combined forces of population and economic growth launched a multipronged attack on the world's tropical forests. Hordes of hungry peasants in Brazil and in African countries tried, with limited success, to hack out farms from the rain forests. Investors sought gold in the rain forests of Brazil. This land was so cheap—it did not need to be bought from the tribal peoples who lived there—that ranchers could chop down vast acreages of rain forest to establish pastures that, because of their poor soil, grew very little grass and produced few and skinny cattle. Some southeast Asian tropical lumber such as teak is of very high quality, but it was so cheap that Japanese companies could make cardboard out of it. Hydroelectric dams flooded vast acreages of tropical forest. It looked like a marauding army had attacked the rain forest. This analogy was more literal than figurative in Brazil, where the government deliberately declared war on its Amazonian "Green Hell," and also moved its capital to a city, Brasilia, newly built in the midst of the drier upland forest.[32]

Very few spots remain intact from this onslaught. Warren Woods in Michigan and Mont St. Hilaire in Québec escaped destruction, for example. A few post oak forests in Oklahoma are almost untouched. The only intact forest in Europe is the Bialowieza Forest, between Poland and Belarus. A grove called Tadasu-no-mori is the only place to see the original forest of the Kyoto area. Shirakami Sanchi, a virgin forest of beech, oak, and walnut, survived owing to its distant location in northern Honshu in Japan.

Many of the temperate North American forests have grown back. Today, many thousands of square miles of deciduous forest cover the eastern states. Forest cover in the twentieth century increased considerably over that of the nineteenth. When we visit these modern forests, in the Smokies, the Appalachians, around the Great Lakes, or in the Rockies, we imagine ourselves to be in pristine woodlands. But, even though these second-growth forests are very beautiful, they are not like the ancient primary forests that preceded them. When a farm or pasture was abandoned, most of the forest woody species returned (although a few, such as the *Franklinia* bush, an American relative of the Chinese camellia, did not). In fact, some tree species such as red maple became even more common than they had been in the original forests.[33] Oak, hickory, maple, and beech trees are abundant in our forests again. But the original forests have vanished. None of us today will ever see the many square miles of magnificent old trees with trunks over a yard in diameter that the pioneers beheld. In their place, in almost every case, is what one writer has called a "forest of sticks."[34]

Some environments have fared even worse than the deciduous forests. The vast tallgrass prairies, once destroyed by plowing, can never come back unless they are deliberately replanted. This is because they depend on a cycle of fires, which cities and farms will not permit. Worst of all are the tropical forests. Many wetlands have been drained and filled in for farming and for human habitation. The tropical forests are now falling as rapidly as, or more rapidly than, the deciduous forests fell a century ago. In many cases, the tropical forests are unable to grow back very well, or at all, after the unsuccessful farms and pastures are abandoned, partly because of the poor soil. Furthermore, tropical seedlings grow well in the small gaps caused by the shifting agriculture of tribal peoples, but not in the large areas cleared by modern agriculture, ranching, and lumbering (see chapter 9).

Not only are the modern forests and fields a pale reflection of what they once were, but they are shattered. What was once a vast uninterrupted natural world is now fragmented. Even the largest state and national parks are small parcels compared with the primeval forest. These parcels are separated by cities, farmlands, and highways. Two of the results of this fragmentation are the edge effect and the loss of gene flow. In small forest parcels, no part of the forest is very far away from the edge, where conditions are too hot, too dry, and too disturbed for many of the forest plant and animals. Because of this edge effect, the amount of land in which deep woods plants and animals can survive is much smaller than the number of square miles that are indicated on a map. Moreover, parcels of forest separated from one another cannot exchange pollen, seeds, or animals. Each parcel consists of small, isolated populations that cannot pool their genetic resources together. This loss of gene flow may cause decreased genetic vigor and reduce the opportunities for evolutionary adaptation.[35]

For these reasons, lots of little Edens do not add up to a big one. This is occurring right at the time when the forests and fields need protection. First, they are facing greater threats from human impact than ever before. Acid rain, caused by air pollution, is not a severe enough problem to directly cause the death of trees; however, it weakens them, making them vulnerable to insect attack and disease. Many of our best forests are dying, according to government officials, as a result of insect pests. These officials are telling half the truth, for it is pollution that makes the insect attacks possible. The improper management of fire is also causing the forests to grow into an unnatural state, thick with dead wood and dense with little trees. By repressing all fires, rather than allowing natural fires to burn, we have created the worst of both worlds: the forests are sick because the fires do not renew them, and the dead wood has piled up so much that if a fire does get started, it becomes a major inferno.

Second, as Earth's climate changes from human impact, the forests and fields need to adjust to these changes by ecological and evolutionary processes. If global warming causes certain national parks to become too hot and dry for the survival of the native species, what are these species supposed to do? In the days before civilization, they could just move to a more suitable location when the climate changed: birds could fly, animals walk, and seeds blow in the wind or hitch a ride on animals. But today, there are roads and fields and cities in the way. The species are trapped in national parks whose climatic conditions are changing. Furthermore, before civilization, the populations of plants in forest and field could adapt to the changes through evolution. Their ability to do so is now reduced because of the loss of genetic diversity. Meanwhile, global warming is occurring ten times faster than it did at the end of the most recent ice age. Because the natural world is fragmented, it cannot meet these challenges, to which it might have been able to respond a few thousand years ago. What this means is that, as the climate continues to change, the surviving fragments of forest and field might vanish. Some scientists see a world in which many of the fields and forests are largely replaced by desert like habitats, populated by the evolutionary descendants of today's weeds and pests.[36]

Who will save the fields and forests of the world? There are a few dedicated environmentalists and a great number of listeners sympathetic to them. Hundreds of millions of people agree that something must be done. But part of the solution will require us to use fewer resources from the earth, and to use these resources more efficiently. We must, in part, deny ourselves some of our luxuries. Denying ourselves for the common good is not a behavior pattern that has commonly evolved among humans, nor is it one that is considered to be "good for the economy." Because we are animals, we will not respond to the environmental crisis with the vigor and urgency that it logically requires. Our heads acknowledge the crisis; our hearts do not, and thus we hesitate. I have often heard major scientists lecturing roomfuls of lesser scientists about the urgency of responding to our environmental crisis, after which the scientists, including myself, filed out of the room and went on with their business. Their lives were essentially unchanged, even though most of them probably agreed with every word that the speakers had uttered.

To anyone who loves the forests and fields, it is sad to see what many centuries, and especially the past century, of human activity have done to the green world of plants. But it is not just a loss to those who feel inspired by a poetic spirit in the woods. It is a loss that threatens everyone who breathes, eats, and drinks—that is, everyone. For plants—big ones like trees and small ones like grasses—are what keep the world alive. One of the purposes of this book is to explain, chapter by chapter, how trees and other plants manage that feat. The list of services that plants provide to our world, and to each of us, is almost inestimable.

Modern technology often contributes to environmental degradation, but it can also allow us to sustain our standard of living without destroying the fields and forests. Environmentally responsible, or "green," technology was an option not available to ancient people. Saving natural habitats and the plants that sustain the world does not have to be at the expense of prosperity. As many economists point out, saving the environment and economic prosperity are not a "zero-sum game" in which one must lose in order for the other to succeed.

Environmental issues command less attention than the many other seemingly more urgent problems facing the nations of the world. Environmentalism is nice, once we have eradicated terrorism and fixed the economy and the educational system, according to many people. Environmental concerns seldom make the list of top priorities, even of the Democratic Party in the United States. The problem is that most people do not understand that all of the other problems of the world are *interrelated with* environmental problems. War and peace, wealth and poverty, education and ignorance, are coupled with the environment. These other issues cannot be solved without addressing environmental problems, nor can environmental problems be solved apart from these others: in a world of war and poverty, the environment cannot be rescued by environmental engineers. As one activist said in response to a person who described himself as not being "into the environment": "Which part of the environment are you not into—the eating part, the drinking part, or the breathing part?"

Nowhere is this more evident than with trees and other plants. Wangari Maathai, a Kenyan woman, studied biology in the United States and earned a Ph.D. in physiology in Kenya. She became active in environmental issues and won the Nobel Peace Prize in 2004 for starting the Green Belt Movement, in which peasants planted literally millions of trees on the slopes of Kenyan mountains. This movement did not begin simply because Wangari Maathai and the poor farmers loved trees, but because they saw that trees were essential to their livelihoods: trees prevented floods, soil erosion, and droughts, and provided essential materials such as firewood. Maathai was not planting trees instead of empowering the poor people of Kenya; she was doing so in order to empower them.[37]

It is one thing to be told and even to believe that we need to save the trees and other plants. It is quite another to actually do it. David Brower pointed out that nobody will try to save what they do not love. He therefore not only championed environmental causes, both during and after his presidency of the Sierra Club, but he also encouraged people to hike and camp and enjoy the mountains and woods.[38] Although the first purpose of this book is to demonstrate the essential value of trees and other plants, its second purpose is to help you appreciate their beauty and wonder. I hope that you support efforts to save energy, save water, save natural areas, and recycle, as well as saving the trees on your own property or as citizens interested in public lands. But I also hope that you go outside, as much as possible, and walk among the trees.

chapter two

PLANTS PUT THE OXYGEN IN THE AIR

If there be a sapling in your hand
When they say to you:
Behold the Messiah!
First plant the sapling, then go out to greet the Messiah.
—TALMUD XXXI

Take a Deep Breath

In order to feel gratitude for the silent, clean, tireless work that plants perform, it is necessary to do no more than to take a deep breath. Better yet, go outside and look at the blue sky, and take a deep breath. Nearly all of the oxygen in the air came from the photosynthesis of land plants, aquatic plants, and other green aquatic organisms. *Photosynthesis* is the process that uses light (*photo-*) to synthesize food molecules. But photosynthesis does much more than that. It is a process so vitally important that it will take three chapters of this book to give it even the briefest overview.

Before photosynthesis became widespread on this planet, there was no oxygen in the atmosphere. Three and a half billion years ago, when Earth cooled off enough for oceans to form, it did not take very long for the first photosynthetic bacteria to appear. They began putting oxygen gas into the oceans and atmosphere. At first, the oxygen that these cells released reacted with minerals such as iron, causing it to rust, and therefore oxygen was trapped in ocean water and sediments and did not accumulate in the air. It took a couple of billion years for oxygen to begin its buildup in the atmosphere. We know that this was happening because the iron that rusted left bands of red deposits in sedimentary rocks. It was not until about 600 million years ago, after the melting of the most recent of three global ice ages, that oxygen gas reached the concentration that it now has in the global atmosphere, about 21 percent by volume. Photosynthetic oxygen also made the diversity of life possible. Every cell uses a molecule called ATP as a source of energy for its metabolic reactions, and the cells of large, active organisms need a lot of ATP. A process called cellular respiration produces a large amount of ATP, and nearly all organisms obtain their ATP from this process. Cellular respiration cannot take place without oxygen. Without oxygen in the air, large organisms could not exist.

New photosynthetic organisms evolved in the oceans, most notably the seaweeds. One group, a certain lineage of green algae, evolved into the first land plants. The first land plants were small and confined to wetlands, but by 350 million years ago there were large trees (which were ancient gigantic relatives of today's relatively small club mosses), ferns, and horsetails. They carried out so much photosynthesis that, according to some researchers, the atmosphere contained even more oxygen 300 million years ago than it does today.[1] Processes such as decomposition of dead plant material would have removed much of this oxygen, but instead of breaking down, much of this dead matter was buried in the bottom of wetlands under sediments. Single-celled photosynthetic organisms accumulated there, and today their remains are oil deposits. The remains of large plants also mounted up, and today their remains are coal. Single-celled organisms in the oceans, when they died, settled to the bottom of the sea, where they were eventually drawn into the crust of the earth by the movement of the ocean floor plates. This process also diminished decomposition, allowing photosynthesis to fill the atmosphere with oxygen.

Planet Earth is, as far as we know, unique in the universe—certainly in the solar system—because of its oxygen atmosphere. The gas giant planets (Jupiter, Saturn, Uranus, and Neptune) are really nothing but atmosphere, though much of it is liquid or frozen. Their atmospheres contain mostly carbon dioxide, methane, ammonia, water, and hydrogen. The atmospheres of Venus and Mars are primarily carbon dioxide. Astronomy has now advanced enough that the light from planets that revolve around other stars can be analyzed in order to determine the composition of their surfaces or of their atmospheres—so far, none have been found with oxygen. What this indicates is that the other planets have no photosynthesis. Because oxygen gas reacts with nitrogen gas and with minerals, it would not be able to persist in the atmosphere of a planet unless it was continually renewed by photosynthesis or some similar process.

How Photosynthesis Produces Oxygen Gas

Plants and other green organisms carry out photosynthesis, a process that is silent and clean, and keeps the earth alive in several important ways. Plants "eat" sunlight and small molecules such as water, carbon dioxide, and soil minerals. They produce carbohydrates such as sugar, and their only waste product is oxygen gas. By making more food than they use, and by producing oxygen, plants are as close as you can get to utterly pure organisms.

The green color of plants is chlorophyll. It is contained within membrane layers of small structures, called chloroplasts, inside of plant cells. Chloroplasts resemble simplified photosynthetic bacteria—which is indeed what they are. Billions of years ago, photosynthetic bacteria moved into a larger cell. The result was a mutually beneficial partnership, in which the bacteria produced food and oxygen, while dining upon the waste products of the host cell. It was such a comfortable relationship that it continued, even when the bacteria lost much of their DNA into the nucleus of the host cell and became modern chloroplasts. Today, chloroplasts have just enough of their own DNA gene expression that we can discern their evolutionary ancestry. These two-in-one cells proliferated into the many forms of photosynthetic organisms on the earth today.

When any molecule absorbs light energy, it becomes warmer. In particular, the light enters the electrons of the molecule, making them more energetic, or, in chemical terms, more excited. The molecule cannot continue absorbing light indefinitely, for eventually the heat will destroy it. The molecule gets rid of the excess energy in several ways. First, it can conduct heat energy into other molecules by making contact with them. This is known as kinetic energy. Second, it can emit light energy as well as heat. All molecules emit low-energy infrared light. A sufficiently energetic molecule can emit visible light, a process known as fluorescence. Third, if the electrons become too excited, they can actually leave the molecule and go to another.

Photosynthesis is the process that captures and uses the electrons that shoot out of chlorophyll molecules in sunlight. The electrons go through what amounts to trillions of tiny electrical wires, or electron transport chains. Electrical wires in buildings are made of copper or iron, whereas electron transport chains in organisms are made of proteins that contain copper or iron. Eventually, the energy from the electrons, and the electrons themselves, enter into carbon dioxide molecules, making them into carbohydrates. These carbohydrates are the source of all the food in the world (see chapter 6). More specifically, the excited electrons from chlorophyll molecules in sunlight are the source of all the food energy in the world.

When a molecule loses electrons, it may fall apart. This is what happens when organic pigments and dyes are left in the sun: the pigments lose their electrons and degrade. Inorganic pigments, such as those that contain iron and copper, are stable in the sunlight. What is it that keeps the chlorophyll molecules from falling apart from the loss of their electrons? These electrons are replaced by other electrons that are released by the splitting of water molecules. Plants absorb massive amounts of water from the soil. About one out of every thousand water molecules absorbed by the roots and pulled up into the leaves is split into oxygen atoms, charged hydrogen atoms, and electrons. The electrons promptly replace those lost by the chlorophyll molecules, allowing the chlorophyll to remain intact. The oxygen atoms unite, and become oxygen molecules (O_2). The oxygen molecules diffuse out of the chloroplasts. On their way out of the plant cell, they may be used by the plant's own respiration, a process that burns food and consumes oxygen, the same process that occurs in our cells. But most of the oxygen molecules diffuse out of the plant cells and into the air. And that is where most of the oxygen in the atmosphere has come from: it is the waste product of the process that replenishes the electrons lost by chlorophyll.

How Much Oxygen Do Plants Produce?

Photosynthetic organisms—the land plants and the plankton in the oceans—are the lungs of the world. The regions of the earth's surface vary tremendously in how much oxygen they produce. Photosynthesis requires water and minerals. The warmer terrestrial zones with greater rainfall and the shallow seas that have the most minerals such as nitrates, phosphates, and iron therefore carry out the most photosynthesis.

A few numbers will give us insight into the prodigious volume of oxygen gas produced by plants. I will use figures published by plant ecologist Christopher Field and colleagues.[2] The net production of either total plant mass or of oxygen is the difference between what the plants produce by photosynthesis (gross production) and what all of the organisms consume by respiration. I make the simplified assumption that the natural habitats of the earth produce about 15 percent as much oxygen gas as they do biomass to produce the rough estimates in table 2.1. Each year, photosynthesis puts more than 17 billion tons of oxygen into the atmosphere. Some regions, such as tropical forests and shallow marine waters, are small but very productive per unit area. The open

TABLE 2.1. Total Net Oxygen Production for the World.

HABITAT TYPE	OXYGEN PRODUCTION (MILLIONS OF METRIC TONS PER YEAR)	PERCENTAGE
Deserts	75	0.5
Tundra	120	0.7
Shrublands	150	1.0
Coniferous forests	465	3.0
Temperate forests	900	5.7
Cultivated land	1,200	7.6
Tropical rain forests	2,670	17.0
Savannas and grasslands	2,880	18.3
Total terrestrial	**8,460**	**53.8**
Oceans	**7,275**	**46.2**
World total	**15,735**	**100.0**

SOURCE: Figures derived from Christopher B. Field, Michael J. Behrenfeld, James T. Randerson, and Paul Falkowski, "Primary Production of the Biosphere: Integrating Terrestrial and Oceanic Components," *Science* 281 (1998): 237–240.

NOTE: 1 metric ton = 1.1 English tons.

oceans are not very productive on an area basis, because the water contains mostly salt and not much fertilizer; but because they are so large, they produce almost as much oxygen as the land plants.

Seventeen billion tons sounds like a lot. But it is small compared with the amount of oxygen that is currently in the atmosphere. At any given time there are about 6.1 quadrillion tons of oxygen gas in the atmosphere. The net contribution of 17.3 billion tons of oxygen that plants put into the air each year is thus only a small fraction of the oxygen. According to this simple calculation, it would take almost 35,100 years for the oxygen to be used up if photosynthesis stopped. No wonder it took photosynthetic microbes such a long time to create the beautiful blue sky almost a billion years ago. But we cannot simply ignore photosynthesis just because the atmosphere contains so much oxygen gas. If we reduced the photosynthetic capacity of the earth's vegetation by half, the atmospheric oxygen content would decrease by 1 percent every 3,500 years. If we expect civilization to continue on this planet at least as long as it has already existed, it is necessary to maintain the current vegetation cover of the earth.

One of the measurements in the table is for cultivated lands, which produce 7.6 percent of the oxygen in the air. Cultivated land is less productive than forests and grasslands. Clearly, saving the forests, and replanting areas that have been deforested, will put more oxygen into the air than planting crops. But even this 7.6 percent is a deceptive figure. Cultivated land requires a tremendous expenditure of energy, which comes from the combustion of fossil fuels. This is especially true of turf. Lawnmowers are among the least efficient combustion devices; they may consume as much oxygen as the grass produces—and that is not counting the trimmers and blowers. Near the beginning of the ecological movement in the United States in the late 1960s, a college campus discouraged students from walking on the grass by placing a sign that indicated that the grassy area produced enough oxygen for two people to breathe. Were it not for lawnmowers and other equipment, this number might be true.

Oxygen levels in the atmosphere actually are decreasing, as combustion and decomposition occur more rapidly than photosynthesis. Climatologist Ralph Keeling has been carefully measuring atmospheric oxygen levels since 1989, and during that time oxygen concentration has declined by 20 parts per million (ppm) each year. At this rate, it would take more than a hundred thousand years to exhaust the atmosphere's oxygen supplies. Ralph Keeling is the son of the late Charles Keeling who showed that carbon dioxide levels have been increasing rapidly in the atmosphere (see chapter 3).

What Would Happen if Photosynthesis Stopped Producing Oxygen?

As the above numbers indicate, the amount of oxygen gas in the atmosphere far exceeds the annual use of oxygen by the respiration of cells, the decomposition of dead biological material, and all of human industry. Also, it is nearly impossible to conceive of a likely situation that could cause all or most photosynthesis to stop. This is why the suffocation of the earth is the least likely environmental danger that we face. But it has happened before. About 250 million years ago, the Permian Extinction brought an end to the Permian period, and to the entire first era of complex life on Earth. It was this event that demonstrated how important plants are in maintaining the atmosphere.[3]

Paleontologist Douglas Erwin called the Permian Extinction the "mother of all extinctions." Approximately 90 percent of species died during and immediately after this event. In contrast, the extinction event that occurred at the end of the Cretaceous period, and which exterminated the dinosaurs and many other species, killed only about 50 percent of the species on Earth. Though the major groups of organisms that had dominated the planet during the Permian (such as seed plants, mollusks, fishes, amphibians, and reptiles) survived, most of the species within these groups did not; the Triassic period, which followed the Permian, began with a vastly impoverished set of species within these groups.

How Did the Permian Extinction Kill So Many Species?

The geological record gives some stark evidence of what happened. First, there are dark-colored sediment deposits at the time of the

Permian Extinction. The dark color is an indicator of the lack of oxygen gas at the time and place the sediments were laid down: the iron is dark green instead of bright red, and undecomposed organic matter darkened it further to black. Such deposits occur today at the bottoms of many ponds and swamps where there is no oxygen. Some of the deposits at the time of the Permian Extinction contain pyrites, an iron-sulfur mineral that forms only in the absence of oxygen. The black slime layer suggests worldwide anoxia, a radical event in Earth's history. In contrast to the 21 percent oxygen in the modern atmosphere, and the possible 35 percent oxygen content during the time when coal swamps were forming, the oxygen content may have dropped as low as 15 percent during the Permian Extinction. According to geologists Raymond Huey and Peter Ward, this would have restricted vertebrates to very limited habitats at low elevations where air pressure was highest.[4] Almost half of the land area of the earth would have been, they claim, uninhabitable by large animals. The black "death bed" layer at the end of the Permian represents between ten thousand and sixty thousand years.

Also at that time, coarse sediments washed down violent rivers. This sometimes included very large boulders that were moved several hundred miles from their point of origin. The structure of river channels suggests as well that rapid erosion was taking place, more rapid than practically anyplace on the earth now, and it was occurring everywhere. This would indicate a worldwide reduction of forest cover (see chapter 7).

Yet another piece of geological evidence is the worldwide shift in oxygen and carbon isotope ratios. Isotopes are versions of atoms that differ in the number of neutrons that they contain, and therefore in weight. Oxygen-18 (^{18}O), for example, is pretty much the same as oxygen-16 (^{16}O), except that it is heavier because of its two extra neutrons. The ratio of oxygen isotopes in geological deposits acts as a permanent record of temperature (see chapter 3). During the Permian Extinction there was a seemingly worldwide change in the ratio of the two isotopes of oxygen, which suggests a massive global warming, averaging 16°F. There was also a global shift in carbon isotope ratios. Plant photosynthesis prefers carbon-12, or ^{12}C, over the heavier carbon-13 (^{13}C) isotope; thus organic matter has a different carbon isotope ratio from inorganic carbon-containing molecules. The global change in carbon isotope ratios during the Permian Extinction, measurable in the geological deposits, suggests a worldwide decrease in plant growth.

The fossils also tell a grim story of what happened during this era. Before the extinction, ferns and conifers were the dominant plants. Fossil pollen was most abundant, as pollen is produced in large quantities, spreads far from the source plant, and has a coat that resists decay. But during the Permian Extinction, fungus spores were abundant. This indicates that the world was literally rotting. Something had killed most of the plants, and mold grew all over them.

During the early period of recovery from the Permian Extinction, plant pollen reappeared, but it was mostly pollen from small plants, such as club mosses, that could grow in recently ravaged landscapes. Furthermore, until recently, there were usually vast swamps of plants somewhere in the world that eventually formed coal. The sediments of the first 20 million years after the Permian Extinction, however, stood out for the absence of coal deposits. The "spike" of fungus spores and the "coal gap" are biotic indicators of a severe devastation of forest growth.

What Caused the Permian Extinction?

The main reasons for the Permian Extinction are probably volcanic eruptions and the release of methane gas. A massive set of volcanic eruptions known as the Siberian Traps occurred at the time. ("Traps" comes from the Swedish word for staircase or steps, referring to the successive layers of lava flow.) These eruptions continued for a few million years. They covered an area of Siberia larger than the entire land mass of Europe. The gases ejected from these eruptions could have caused worldwide devastation. Sulfur dioxide (SO_2) from volcanic gas reacts with water to produce sulfuric acid, a component of acid rain. Severe acid rain may have killed much of the vegetation on land and photosynthetic organisms in the oceans, causing the collapse of oxygen production and of food chains in both. Other large volcanic eruptions during Earth's history were not associated with mass extinctions, but the Siberian Traps eruptions may have had a global impact if they ejected more sulfur than typically occurs. The lava from these eruptions was rich in sulfur minerals, suggesting that they released even more sulfur dioxide than most volcanic eruptions observed today. Volcanoes also eject large amounts of carbon dioxide, which is a principal cause of the greenhouse effect (see chapter 3), and could have been the cause of the global warming that is recorded in the oxygen isotope ratios mentioned above. The resulting death of plants could have brought about the worldwide plunge in oxygen levels.

Along with volcanic eruptions, the global warming from the volcanic gas may have caused yet another set of catastrophes to occur: gigantic global burps. Deep beneath the sediments just beyond the continental shelves, especially in polar regions, there are today (and may have been during the Permian) large deposits of methane hydrate, which is an ice-like combination of water and natural gas. Bacteria-like cells produce the methane, which is hydrated by water pressure and cold temperature. If the ocean waters became warm, the methane might evaporate explosively, bubbling quickly through the ocean. The gas would probably have become carbon dioxide by the time it reached the surface of the ocean, but such a methane eruption would have emitted a gigantic amount of carbon dioxide into the atmosphere. There is evidence that such "burps" have occurred in the geologically recent past: such an eruption 55 million years ago may have caused a global warming of about 5–7°F over a ten thousand–year period. By enhancing the global warming that was already going on, these methane eruptions could have started a positive feedback loop in which more global warming caused the release of even more methane. Methane reacts with oxygen, and this would have worsened the already dire problem of oxygen depletion.

At the end of the Permian, fossil evidence of plants (for example, pollen) nearly vanished, and so did oxygen. There is no need to speculate that life on Earth is strongly dependent on the work of plants: the "experiment" has in fact been done. This is part of the bad news. Another part of the bad news is that events in Earth's history—as, in this case, normal volcanic activities—can have a severe impact on life. Evidence has been found for a possible asteroid collision with Earth about the time of the Permian Extinction. But the processes that led to the Permian Extinction appear to have been well under way before this asteroid hit. It does not take an asteroid to kill almost everything off.

The good news is that the earth can recover from an almost total extinction event. No matter what humans do to the earth, and probably no matter what happens to it from other causes, except for the final explosion of the sun several billion years from now, the earth will recover. But this good news is also bad. The recovery would take forever, from the viewpoint of human history. Moderate recovery from the Permian Extinction to a normally functioning natural world took at least 10 million years, and the full recovery of the number of species, as new ones evolved in place of the ones that had become extinct, took 100 million years. The human economy is dependent on the continued stability and smooth operation of natural systems. For humans, even a slight change in global temperature would spell agricultural and economic disaster. Because civilization relies on the exploitation of natural resources, even the slightest interruption of these resources would send humans, very few of whom know how to survive in the wild, into a desperate tailspin. Humans would probably not have survived the Permian Extinction, or a destruction of plant life even one-tenth as severe as that event. The loss of oxygen would be one of the least of the problems to emerge from such a catastrophe.

Photosynthetic Oxygen Protects the Earth from UV Radiation

The spread of photosynthesis across the land and water surface of the earth produced the oxygen in the atmosphere, but it was also the product of it. It is the oxygen in the atmosphere that shields the surface of the earth from deadly levels of ultraviolet (UV) radiation from the sun. UV radiation occasionally splits oxygen molecules into oxygen atoms, which almost immediately react with another atom or molecule—and if an oxygen atom fuses with an oxygen molecule, the result is a molecule that consists of three oxygen atoms: ozone. Ozone absorbs UV radiation very effectively. Therefore ozone is both a product of, and protection against, UV radiation. Life forms could not emerge from the depths of the sea until oxygen had begun to accumulate in the atmosphere and an ozone layer had formed. Bacteria made food from the energy in hydrogen sulfide from volcanic vents, and photosynthetic bacteria used the dim light from volcanic vents until about a billion years ago.[5]

UV radiation is dangerous primarily because it damages sensitive tissues of organisms. This includes the retinas of animals' eyes, but the most noticeable danger is skin cancer. DNA absorbs UV radiation and can experience mutations as a result. In many cases, these mutations can result in cancer. The "ozone layer" protects the earth from UV radiation. The ozone is dispersed at low concentration through a couple of thousand feet of atmosphere. If the ozone were consolidated into a single, pure layer, it would only be about a half-inch thick. This layer is not much, but it protects us from intense sunlight.

Industrial chemicals, most famously the chlorofluorocarbons (CFCs), have accumulated in the upper atmosphere in only very small quantities. And yet, under the very cold conditions over Antarctica, the chlorine atoms from CFCs began a chain reaction in which each chlorine atom could destroy hundreds of ozone molecules. The result was a striking "ozone hole" over Antarctica, of which now nearly every educated citizen of the planet knows. A smaller, less depleted "hole" also formed over the Arctic (which, being at sea level, is not nearly as cold during the boreal winter as Antarctica during the austral winter), and ozone levels were measurably, though not dangerously, depleted in the entire atmosphere of the earth. Quick action on the part of governments, and cooperation from the industries that produced CFCs, resulted in an amazingly effective international agreement, the Montreal Protocol of 1987, which has vastly reduced the release of CFCs into the atmosphere, allowing the hope that the ozone hole will eventually heal. There is even evidence that the low atmospheric oxygen levels of the Permian Extinction were associated with a weakening or even a destruction of the ozone layer.

The production of oxygen is one of the most important ways in which plants keep Earth alive. Fortunately, the oxygen that they have already produced will be enough to sustain life on Earth even if there is a temporary, human-induced reduction in the capacity of plants to carry out photosynthesis. The same cannot be said, however, of other human threats to this process: the photosynthetic removal of carbon dioxide from the air and the production of food—the topics of later chapters.

chapter three

GREENHOUSE EARTH
PLANTS HELP TO KEEP THE EARTH FROM OVERHEATING

And he also said to the multitudes, "When you see a cloud rising in the west, you say at once, 'A shower is coming'; and so it happens. And when you see the south wind blowing, you say, 'There will be scorching heat'; and it happens. You hypocrites! You know how to interpret the appearance of earth and sky; but why do you not know how to interpret the present time?"
—LUKE 12:54–56

The earth is getting warmer, and humans are the principal cause. Humans have increased the amount of carbon dioxide (CO_2) gas in the atmosphere, and this gas retains heat in the atmosphere. Along with global warming have come other changes in climate, such as an increase in the intensity of storms. Consensus of scientists on these points has grown over recent decades and is now as strong as any scientific consensus can be. Plants can help to reduce global warming. Considering how important a threat global warming is to the future of humankind, we need all the help we can get in preventing it, and that includes help from plants.

The Earth Is Getting Warmer

Weather Is Not Climate

A fierce, frigid wind rammed down Washington Avenue between America's Center and the Renaissance Grand Hotel in downtown

St. Louis in February 2006. It was almost strong enough to blow away the thousands of scientists, including me, who were attending the annual meeting of the American Association for the Advancement of Science. Many of them had come to discuss global warming. A global warming skeptic, if there were any present, would have said: "So where is global warming now?" This was exactly the reaction that some people had to the April 2007 "Step It Up" campaign, started by science writer Bill McKibben to focus political attention on global warming. Step It Up 2007 took place just as a frigid winter storm barreled down on the whole United States. These and many other weather events seem to discredit the idea that the earth is becoming warmer.

On the other hand, many weather phenomena do strongly support that proposition. In the summer of 2006, I returned to the town in which I had grown up—Lindsay, in the San Joaquin Valley of California. When I was a child, the temperature exceeded 100°F on many days each summer. But in July 2006, the air almost made me faint, with a temperature exceeding 115°F. I was convinced that this was the result of global warming. Some long-term observations made by individuals also seem to prove that the earth is getting warmer. In December 1983, I stood on the shore of Lake Oologah in northeastern Oklahoma. A thick layer of ice covered the lake, and the water underneath the ice lurched and mumbled eerily. Since that time, the winters have been warm enough that the lake has not frozen. My mother remembers abundant snow and ice in the Oklahoma winters of the 1920s.

But none of these events are, by themselves, evidence for or against global warming. They are observations of weather, very limited in space and in time. Weather is not climate. The same is true of other weather events such as hurricanes. Scientists predict that global warming will cause hurricanes to be more severe. Many people believe that the hurricane season of 2005 proved this. Hurricane Katrina caused unprecedented damage and disruption in the United States, and Hurricane Wilma had the strongest wind velocity ever recorded in such a storm. That season also held the record for the number of hurricanes. But then the 2006 hurricane season was mild, and the global warming skeptics claimed that, after all, nothing unusual was happening. Then, when

Hurricane Dean became the first Category 5 hurricane in a long time to make landfall, the defenders of global warming again felt vindicated. Many people were also convinced that global warming was the cause of the violent burst of tornados (more than seventy of them) in the American South in February 2008. Tornadoes usually occur in the spring and summer. Although global warming is causing spring to begin earlier than in previous centuries (see below), we cannot attribute every unusual weather event to global warming. Weather is not climate.

The Earth Is Warmer Than It Has Been in a Half Million Years

In order to determine whether global climate is becoming warmer, it is necessary to determine temperature patterns over extended periods of time, and over the whole earth. Climate can be thought of as a long-term average of weather conditions. One hot year does not prove that global climate is becoming warmer. However, since scientists began recording temperatures in many parts of the world about 1850, global average temperatures have significantly increased. Furthermore, eleven of the twelve hottest years have occurred just in the past twelve years (fig. 3.1).[1] According to the NASA Goddard Institute for Space Studies, 2007 tied with 1998 as the warmest year in the past century. North Atlantic Ocean temperatures oscillate between cool and warm over the course of decades, but they also show a pattern of recent warming.[2] This is the sort of evidence that is required to show that global climate has become warmer since the beginning of the Industrial Revolution and particularly in the past half-century—and the evidence has been found. The earth has not only become warmer, but abruptly warmer.

Even though weather is not climate, we can assign a probability that global warming has contributed to particular weather events. Climate scientists can use their computer models to calculate the probability that certain climatic events would have happened without the extra carbon dioxide that has been released into the atmosphere since the beginning of the Industrial Revolution. These models take into account natural factors, such as the movement of the earth and the El Niño

FIGURE 3.1. Global average temperatures have increased during the past century. This has been documented by actual thermometer measurements from many parts of the world. The record-breaking world temperatures of the past three decades far exceed even the heat wave that occurred during the North American Dust Bowl of the 1930s. The graphed line represents five-year running averages. Figure by the author, based on data from the NASA Goddard Institute for Space Studies (GISS).

Southern Oscillation of ocean currents. This approach allows scientists to state that about half of the 2006 North American heat wave could be attributed to global warming, and half to natural variability.[3] We can therefore in retrospect say that global warming has contributed to the overall pattern of weather events.

A few scientists deny either that global temperatures are increasing or that human activities are an important cause. They have not, however, verified their claims in the scientific literature. Their claims are largely limited to their own Web sites or to books published by organizations that have received funding from corporations that have a vested interest in the continued burning of fossil fuels. One example is climate scientist Sherwood Idso, who with his sons Craig and Keith has posted long-term U.S. government temperature records on their anti–global warming Web site.[4] They claim that the climate has not gotten warmer in the past century and that it has actually gotten cooler in some selected locations. However, they omit the much larger number of places in which the weather has gotten warmer. Their Web site invites visitors to view the data for themselves, so I decided to do so. I went to the "Temperature Record of the Week" for each of the archived issues in the first half of 2007. Of records for seventeen locations, each in a different state in the United States, only five showed a decrease over the past century, and four of those had a decrease of 1°F or less. Two of these decreases were owing to cooler temperatures between 1960 and 1980, after which the temperatures had begun to increase again. The Web site uses graphs in which a straight line is imposed upon the data, which tends to obscure the temperature increases since 1980. One of the records showed very little change. Eleven of the records showed temperature increases, five of them at least 2°F. The average temperature change, from the data posted on this site, was an increase of about half a degree. So the Idsos' data illustrate the same overall warming trends as do global average temperatures, and it is safe to say that this Web site does not present evidence that supports the authors' claim.

Even though few scientists, and no published data, call global warming into question, journalists and the public have tended to perceive the claims of "global warming skeptics" as potentially valid. Climatologist Naomi Oreskes surveyed 928 scientific articles that dealt in some way with global climate change. None of these articles denied that global warming was occurring. However, when she surveyed 636 articles published in popular media, 53 percent of them gave credibility to the claims of those who deny global warming.[5] Apparently the public has been misled about the truth of global warming.

But a century is not a very long time. The important question is not whether the climate is getting warmer, but whether this increase in global temperature is the result of human activity. If humans are partly or largely responsible for global warming, it must have occurred since the beginning of the Industrial Revolution and accelerated as industrial pollution has spread throughout the world. If this is the case, the recent increase in temperature must be greater than any that occurred in the past one or two millennia. Therefore scientists have investigated temperature patterns further into the past than is possible with just the "instrumental" (thermometer) data. Scientists must use other methods to estimate temperatures before that time. Several of these "proxy" methods have been developed.

One example of a proxy method is the measurement of tree rings. Each year, trees add another layer of wood just underneath the bark. The layer is visible as a ring because the wood produced in the spring is more porous, therefore lighter in color, than wood produced in late summer. The age of the tree can be accurately determined by counting the rings all the way to the center of the trunk. During years with favorable weather conditions, trees produce a thick ring of wood; when the weather is poor, the ring is thinner. Good and bad weather result from temperature and moisture conditions. For trees in the north, good years are warm and bad years are cool; for trees in the south, good years are wet and bad years are dry. The barcode pattern of thick and thin layers of wood represents a record of good and bad years, extending back into the past, as long as the tree has been alive. Some trees, such as the bristlecone pines of California, have been alive and accumulating layers of wood for more than four thousand years (fig. 3.2). Dendrochronologists, scientists who study tree rings, do not limit themselves to the study of trees that are alive right now. If they find a well-preserved log, they can line up the barcode pattern of thick and thin rings in the log with those in a living tree, allowing them to determine what year the tree died. They can then extend the tree ring record back to the year the dead tree had begun growing. By this method, forest scientist Edward Schulman was able to determine that a certain bristlecone pine log at the Ancient Bristlecone Pine Reserve in California, now called the Schulman Log, began growing about 1530 BCE and died in 1676 CE at the age of 3,206 years. Tree rings in very long-lived trees are more difficult to interpret than those in trees that grow faster and have shorter life spans. But the climate record from tree rings is fairly reliable for the past thousand years, especially when verified by other proxy methods, such as the chemistry of the annual layers of sediment on continental shelves.

The conclusion of these proxy methods is that the global climate today is hotter than it has been at any time in the past thousand years,

FIGURE 3.2. The very tiny growth rings of bristlecone pines (*Pinus longaeva*) in the mountains of eastern California contain a record of ancient climatic patterns. Photograph by the author.

and the current warming trend is faster than any warming trend during that period. Several different proxy reconstructions of past temperatures have been published by scientists, and all are in close agreement with this major conclusion. The most famous of these graphs is climate scientist Michael Mann's "hockey stick model," so called because a long period of relatively stable temperature (the hockey handle) has given way to a sudden warming as brief and abrupt as the end of a hockey stick (fig. 3.3).[6] Since its publication in 1998, this graph has withstood criticisms, one of which involves the so-called Medieval Warm Period and Little Ice Age. Over the past millennium in Europe, the average temperature has fluctuated. Temperatures were higher during the Medieval Warm Period in the fourteenth century and thereafter became cooler, producing what has been named the Little Ice Age. Some observers noted that Mann's graph does not show these two events. It turns out, however, that these phenomena may have been limited to Europe or to the northern hemisphere, rather than global in extent. One reason for the fame of this graph is that it so angered a

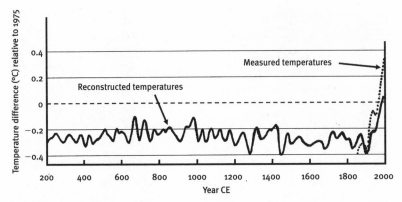

FIGURE 3.3. These estimates of global land surface temperature from 200 CE to the present were produced from indirect measurements of temperature, such as tree rings, by Michael Mann and associates (see note 6). The dotted line for the past century and a half is based on actual temperature measurements, as presented in figure 3.1. Figure by the author.

member of Congress (Joe Barton, a Republican from Texas) that he began an investigation into the Mann's competence, a move that was widely recognized as an attempt to suppress the evidence for global warming. In 2006, a panel of the National Academy of Sciences confirmed Mann's data, although it noted that the estimates of temperatures of the past four centuries are more reliable than the estimates of the older dates.[7] The Intergovernmental Panel on Climate Change (IPCC) is a group of scientists and government representatives from more than a hundred nations, including the United States, brought together by the United Nations Environmental Programme (UNEP) and the World Meteorological Organization. Their reports represent as close to an international consensus on global warming as we can hope to expect. A 2007 report from the IPCC contained a figure that shows essentially the same pattern as the original graph by Mann.[8]

But even a millennium is not very long in terms of human existence, not to mention that of the planet. Climate change has been occurring throughout Earth's history. Among the most dramatic changes have been the ice ages. For about the past two million years (since the beginning of the Pleistocene epoch), the northern hemisphere has undergone a series of glacial cycles. During every one, cold temperatures have caused glaciers to accumulate and spread southward from the Arctic Ocean (glaciation), and warm temperatures have caused them to melt and retreat (interglacial periods). Each cycle has lasted about a hundred thousand years; therefore since the beginning of the Pleistocene two million years ago about twenty glacial cycles have occurred. Each glacial advance effaces much of the evidence of earlier cycles, but direct evidence, such as grooves scratched in rocks and piles of rubble left by melting glaciers, can be found for at least four. The most recent glaciation ended about fourteen thousand years ago, and the glaciers began to melt rapidly, ushering in the most recent interglacial period, which is still going on. About seven thousand years ago, peak interglacial temperatures occurred, producing warm and dry conditions across much of North America. After that time, temperatures decreased, as the northern hemisphere headed toward the next glaciation.

This raises the question of whether the recent global warming is merely part of a natural cycle of warming. There are at least three reasons why this is not the case. First, global temperatures were decreasing after the interglacial temperature maximum about seven thousand years ago; recent warming represents a departure from this trend. Second, global warming since 1850 has occurred faster than previous periods of global warming. Finally, global temperatures appear to be higher than at any time during any of the most recent glacial cycles.

The evidence for these three claims comes primarily from the study of ice cores. In some parts of the world, snow has accumulated for many thousands of years. As new snow piles up, the old snow underneath it is compressed into ice. For example, ice has amassed in glaciers in the Andes Mountains of South America and on Mount Kilimanjaro in Africa. The longest accumulation of snow has occurred in the thick ice caps of Greenland and Antarctica. The Antarctic ice can be two miles thick and heavy enough to press the mountainous continent of Antarctica down below sea level. Scientists can (with considerable difficulty) drill down into this ice and remove cylinders that (in the case of the Vostok ice core from Antarctica) date back more than four hundred thousand years. Like tree rings, ice builds up in easily visible layers, caused by annual cycles of snow. The age of a layer of ice, therefore, can be determined by counting the layers downward from the top (fig. 3.4).

As with tree rings, the layers reveal much more than just the age of the ring. First, scientists can determine whether the year that the ice layer formed was relatively warm or cool. Each water molecule contains one oxygen atom. Most oxygen atoms have an atomic weight of 16, resulting in a molecule of $H_2{}^{16}O$. A relatively small number of oxygen atoms, however, are heavier isotopes, with an atomic weight of 18, producing $H_2{}^{18}O$ molecules. The lighter water molecules evaporate from the ocean surface more readily than the heavier molecules and accumulate in the ice on land. A greater proportion of light water in the ice layer indicates a cool year in which a greater amount of ice built up. Second, there are bubbles of air trapped in the ice (also visible in fig. 3.4). Scientists can remove the gas from the ice layer and measure the amounts of each kind of gas that was in the atmosphere at the time the layer was formed. The importance of this procedure will become obvious a little later in this chapter.

FIGURE 3.4. An ice core, showing annual layers of ice with air bubbles, on display at the American Museum of Natural History. Photograph by the author.

Using this information, scientists can reconstruct a record of global average temperatures reaching back as far in the past as the ice layers will allow. An analysis of the changes in global temperature derived from the Vostok ice core from Antarctica clearly reveals the four most recent glacial cycles, each with glacial and interglacial periods (fig. 3.5).[9] Global temperatures have not yet increased beyond those of previous interglacial periods, but, as previous graphs have shown, that increase is now occurring rapidly. Global warming is nothing new in Earth's history; but the current rate of global warming is ten times as rapid as the warming that occurred at the end of the most recent glaciation.

There is other evidence that global temperatures have been increasing during the recent centuries and millennia. The most famous evidence is that, all around the world, ice that has accumulated for thousands of years is beginning to melt. In Antarctica, shelves of sea ice have been breaking off and forming icebergs. Because Antarctica is the coldest place on earth, and because it has so much ice, scientists had expected

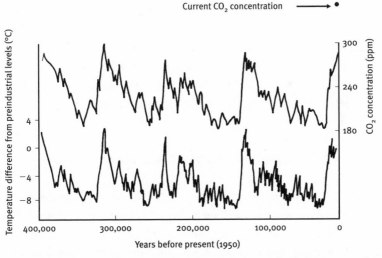

FIGURE 3.5. The estimates of temperature (based on oxygen isotopes), relative to present-day surface temperatures, and carbon dioxide (from the direct measurements of gas in the ice) for the past 420,000 years, based on ice cores from Vostok, Antarctica. The rapid increase in carbon dioxide during the past two centuries has been added to the graph as a single point. Figure by the author.

that global warming would melt Antarctic ice slowly and gradually. Much to their surprise, they found that it is melting more rapidly than they could have imagined. The Larsen B ice shelf in Antarctica, the size of Rhode Island, rather than gradually melting, collapsed over the course of two months in 2002, most of the cave-in occurring during two eventful days. A portion of the Wilkins ice shelf collapsed in March 2008. Meanwhile, in Greenland, sheets of land ice are melting and retreating. As in Antarctica, Greenland ice is thick, and scientists expected it to melt slowly. They were surprised to discover in 2006 that water from glacial melt leaks down to the bottom of the glacier and lubricates it, allowing it to move relatively rapidly toward the ocean.[10] In addition, historical records and comparisons of photographs prove that glaciers all over the world are melting away. The glaciers in the Andes and on Kilimanjaro may be gone in just a few decades. Finally, the extent and thickness of the sea ice in the Arctic Ocean are decreasing. Just in the past half-century, the Arctic ice cap has become 20 percent smaller and several yards thinner. Given that some of this ice has built up over hundreds of millennia, the conclusion is unavoidable that global temperatures are heating up faster than at any time in the past four hundred thousand years, and perhaps more than at any time since the beginning of the Pleistocene. These temperature increases have been occurring even though the oceans have absorbed much of the heat that would otherwise have contributed to global atmospheric warming.[11]

Ocean Currents and Sunspots Affect Global Temperature Patterns

One of the most important factors in global climate patterns is the movement of ocean currents. They redistribute heat from tropical zones to the high latitudes throughout the world. One of the most important patterns of water movement has been called the "global conveyor" current. (It is a general pattern of water movement rather than a discrete current.) Part of this system is the Gulf Stream, in which warm surface water moves from the tropical Atlantic up toward western Europe. This warm water makes the air warmer in western Europe than at the same latitudes in Canada. From there, the warm water descends abruptly to the bottom of the North Atlantic, and moves down the east coast of South America. It then flows past Antarctica. The water, now cool, flows northward to the tropics, becoming warmer as it moves north toward Japan and turns east, bringing warm weather to the vicinity of British Columbia and Washington. The water moves back toward Australia, south of Asia and Africa, and returns to the tropical Atlantic Ocean where it replenishes the Gulf Stream. Another important pattern is the El Niño Southern Oscillation (ENSO). When warm water flows from tropical Asia toward South America, surface waters are warm and El Niño storms are strong and frequent in South America. When cooler water, drawn from the ocean depths off South America, flow toward tropical Asia, drier La Niña conditions prevail in South America. Nobody knows how these ocean currents interact. Is the global conveyor current a cause, or an effect, of climate change—or both? The shifting effects of ocean currents on global temperature patterns is one of the reasons that global warming has not occurred uniformly during the past century.[12]

The fact that ocean currents make a tremendous difference in global temperature patterns is demonstrated by something that happened about three million years ago. Before the isthmus of Panama arose at that time, ocean water moved unimpeded through the gap between North and South America on its way to the Arctic Ocean. This strong current delivered enough warmth to the arctic regions that broad-leaved deciduous trees were able to grow nearly to the North Pole. The formation of the isthmus of Panama considerably weakened this current, so that deciduous trees now form extensive forests only about halfway to the North Pole. Scientists have concluded that this cannot be the entire reason that the North Pole was much warmer in the past than it is today, but the decrease in ocean currents was probably an important factor.

If anything slowed or stopped the Gulf Stream, the weather in northern Europe could become much colder. This apparently happened about 12,800 years ago, right when the glaciers of the most recent ice age were melting and the earth was becoming warmer. Very quickly, the warming trend reversed, and temperatures as cold as those of the glacial period returned. This massive cold snap is called the "Younger Dryas." *Dryas* is the name of an arctic wildflower, whose range had retreated northward as the glaciers melted, but that quickly returned to Europe and North America when the temperatures became frigid again. The cold snap was apparently caused by a flood of freshwater that entered the North Atlantic. The collapse of a glacier released a huge amount of cold water through the St. Lawrence Seaway into the North Atlantic Ocean. The cold freshwater was lighter than the saltwater of the ocean and remained near the surface of the North Atlantic. Then the warm Gulf Stream waters apparently sank long before they reached Europe. Europe, no longer receiving Gulf Stream warmth, experienced a rapid chilling. Nobody knows what caused this sudden glacial melt.

Could the North Atlantic again be flooded with cold freshwater? The melting of Greenland ice could cause something like this to occur. Therefore the effects of global warming may, paradoxically, include a temporary, sudden cooling in Europe. This is, however, one of the predictions that has the least support among climate scientists.

Sunspots also appear to affect global temperature. For example, there was very little sunspot activity during the Little Ice Age. As is the case with ocean currents, variations in sunspot activity appear to superimpose global temperature variations on an overall warming trend.

What Will Happen as the Earth Becomes Warmer?

It is clear that the earth is becoming warmer. How much warmer will it become in the future? In order to predict future temperatures, scientists run complex "general circulation models" on some of the world's most powerful computers. These models make, and integrate, predictions for each of many grid squares into which the earth's surface is divided. Different computer models are used at the Goddard Institute for Space Studies in New York, Lawrence Livermore Laboratories in California, the Hadley Centre at the University of East Anglia in England, and the Max Planck Institute in Germany. Although they are all different models, they yield very similar results. And there has been at least one opportunity to test the models for reliability. When the volcano Pinatubo erupted in the Philippines in 1991, it spewed particulate matter into the upper atmosphere. Using the Goddard Institute program, climatologist James Hansen predicted that global temperature would decrease by about half a degree F as a result. His prediction turned out to be very accurate.

In the sections below, I consider some of the consequences of global warming.

Global Warming Is Raising the Sea Level

Many of us grew up thinking that "sea level" never changed. The signs outside American cities proclaim how many feet above sea level they are. However, during the most recent glaciation, the sea level was almost 300 feet lower than it is today, because enormous amounts of water were locked up in glaciers. Then the glaciers melted and the ocean levels rose. Scientific measurements indicate that ocean levels are rising by about a quarter of an inch per year. The sea level appears to be rising at a faster rate than in the past, as a result of global warming. There are two reasons for this. First, when sea ice melts, the additional water does not alter the ocean level. But when global warming causes ice on land to melt, the water drains into the ocean and raises its level. Water from melting glaciers, especially in Antarctica and Greenland, is already causing the sea level to rise. If all of the ice melted, which may occur in a few centuries, the ocean levels would rise several feet. Second, as global temperature increases, the oceans become warmer, causing the water to expand.

If the Greenland ice cap melted, or just fragmented and fell into the ocean as icebergs, the ocean level would rise by about 20 feet. The West Antarctic ice sheet, if it slid into the ocean, would cause the oceans to rise another 20 feet. (The East Antarctic ice sheet is much larger but is more stable and unlikely to collapse.) This rise would be enough to flood the major coastal cities of the world, such as New York, Boston, Washington, Los Angeles, San Francisco, Seattle, Shanghai, Tokyo, Bangkok, Bombay, London, and Marseille. The ocean levels would rise slowly enough that people would have time to move, but the expense of moving and rebuilding—or, alternatively, of building sea walls—is nearly incalculable. The map of the world would be redrawn, including a Pacific Ocean in which several island nations have vanished, and a United States with most of Florida missing. The sudden influx of glacial meltwater into the North Atlantic at the start of the Younger Dryas raised ocean levels by more than 30 feet.

Global Warming Is Causing More Rainfall in Coastal Areas and Droughts Inland

Warm air can hold more moisture than cool air. As a result, when temperatures rise, more water evaporates from the oceans, most of which falls in coastal areas. The average amount of rainfall in the world has already increased by 20 percent in coastal regions. But warm air can also cause more water to evaporate from plants and soil. Therefore, away from the oceans, in continental areas, rainfall could decrease. These continental areas include most of the agricultural region of the United States, where droughts are becoming more frequent. A study released in 2007 predicts that in the American Southwest droughts will become the norm rather than the exception.[13] This region depends on water from snowmelt, and if snowfall decreases, large areas may experience droughts severe enough that crops cannot be irrigated. Meanwhile, in the middle of Africa, droughts may become more frequent and intense. Moreover, the year 2008 was the seventh year of one of Australia's worst droughts. The drier conditions will undoubtedly cause more frequent, and more severe, wildfires.[14] Fires burned large areas of forest worldwide in the summer of 2005, not only in the tropics but in Alaska as well.

Global Warming Is Causing Stronger Storms

Even though 2005 had more hurricanes in the Atlantic and more typhoons and cyclones in the Pacific than any previously recorded year, the historical record of storms indicates that the frequency of hurricanes is not increasing. Every few decades, a period of either more or fewer hurricanes begins. However, recent hurricanes and other storms have been more severe than those in the past.[15] In India, for example, there has been an increase in heavy storms, but not in the total number of storms, over the past half-century.[16] The intensity of storms is influenced by the warmth of the surface waters in the North Atlantic Ocean; warmer surface waters cause more evaporation and stronger wind, both of which contribute to creating intense storms. Hurricanes Katrina (in 2005) and Dean (in 2007) may therefore give us a very realistic glimpse into the kind of damage that will become commonplace in future decades, as global warming continues. In a warmer world, strong storms can also form in places where they have not previously occurred. The first recorded South Atlantic hurricane occurred in 2004.

Global Warming Is Causing Changes in Plants and Animals

Global warming is already having measurable effects on plants. Plant species are growing closer to the poles and further up on mountains as a result of global warming. Antarctica has only two native species of higher plants, one of which is the grass *Deschampsia antarctica*. It used to form only widely spaced tussocks in protected crevices along the coast, but Antarctica now has its first grass meadows. A major indicator of global warming is that buds of deciduous trees in the northern continents are opening sooner now than even a few decades ago—spring is coming earlier. A few isolated instances stand out. Perhaps the most famous example, and most noticed by the United States federal government in Washington, D.C., was when the cherry trees bloomed along the Potomac in January 2007, far sooner than their usual budburst time.

But scientists do not rely on isolated examples. Botanist A. J. Miller-Rushing and associates compared pressed plant specimens and photographs from the nineteenth and early twentieth centuries with recent specimens and photographs.[17] Their analysis shows that trees are leafing earlier, and herbaceous plants emerging earlier, in the very same locations—about eleven days earlier per century. Miller-Rushing and colleagues used 285 photographs and numerous preserved specimens. This pattern is confirmed by satellite imagery, which indicates that deciduous forests are becoming green earlier in the spring than in previous decades. Another study, published by climatologist Xiaoyang Zhang and associates, shows that lilacs planted all over North America have been flowering earlier in the spring. Interestingly, some of the southern lilacs have been blooming later rather than earlier, because in those southern locations the winters have not been cold enough to allow the biochemical breakdown of budburst inhibitors.[18]

In addition to changing the seasonal patterns of plant growth, global warming is also causing changes in where the plants live. Ecologist Camille Parmesan has reported the results of a survey of 1,700 plant species. On average, these plant species migrated northward about 4 miles per decade, and migrated higher up in the mountains by about 24 feet per decade. She also found that, on average, spring budburst occurred about two days earlier per decade.[19] Daniel W. McKenney of the Canadian Forest Service and associates studied 130 tree species and have estimated that most of them will experience severe reductions in geographical range even if they manage to migrate hundreds of miles northward.[20] In Britain, 385 species of plants flowered an average of four and a half days earlier over the course of a decade.[21] In Spain, high-altitude grasslands are being replaced by lower-altitude shrublands.[22]

Animals are also responding to climatic changes. Armadillos have been expanding from their native habitats in the South, as far north as Illinois. Several bird species have moved their breeding ranges northward. Over the course of twenty years, British birds have shifted their ranges about 12 miles northward, and British birds are laying their eggs sooner.[23] Camille Parmesan studied not only plants but also nonmigratory European butterflies, most of which shifted their range northward from between 22 to 148 miles during the twentieth century.[24] Global warming has caused changes in migration patterns of some birds. A species called the European blackcap breeds in Germany and has traditionally migrated to Spain and Portugal to overwinter. Recently, some flocks of blackcaps have spent the winter in England instead. The flocks return from England earlier than the flocks from Portugal, with the result that the England-wintering birds mate earlier and leave more offspring.[25]

One reason that these shifts in range and seasonality are important is that they may result in extinctions. The plants and animals that already live at the tops of mountains, such as the alpine tundra of the Rocky Mountains and Sierra Nevada, and the cloud forests of Central America, have no place to go if conditions become warmer or drier. The alpine tundra, which consists of small grasses, sedges, and wildflowers, survives only because conditions are too cold for trees to grow over them and shade them (chapter 8). Once the subalpine spruce and pine forests begin to grow to the very tops of the mountains, the tundra plants and animals will face extinction. The cloud forests of Central America persist because they are inside of clouds and receive abundant water from mists. Warmer temperatures are causing the cloud levels to rise, leaving the cloud forests high and dry. The golden toad has already become extinct because of the drier conditions in the cloud forests. Ornithologist Çağan Şekercioğlu and associates have predicted the extinction of between one and five hundred bird species for each 1°F increase in global temperature.[26]

Global warming can also disrupt ecological relationships between species. Warmer temperatures may give insects a longer reproductive season and allow them to spread rapidly on their host plants. (This includes the insect pests of crop plants.) Warmer temperatures are already allowing outbreaks of bark beetles to destroy millions of spruces, lodgepole pines, and piñon pines.

Other disrupted ecological relationships result when global warming causes some organisms to become active earlier in the spring than do others. One example is the relationship between flowering plants and their insect pollinators. Most wild plant species, other than conifers, grasses, and sedges, depend on insects for pollination. If warmer temperatures change the seasonal life cycles of the plants and the pollinators in different ways, the flowers may open when the pollinators are not present. As a result, the pollinators may find no food, and the flowers will not be pollinated. Another example is the disruption of the food chain. Herbivorous animals may respond more quickly to the warming climate than do the plants that they eat, with the result that the animals may hatch and find little or nothing to eat.[27] Or, if warmer springs cause caterpillars to hatch before the baby birds that eat them, the nestlings may not have enough to eat.[28] Within bird populations, those individuals that could best adjust the timing of their reproduction have been more successful than those that could not alter their reproduction to adjust to the earlier arrival of spring.[29]

Emerging from dormancy earlier in the spring may even cause direct harm to organisms. This could occur if an extended warm period in the winter awakened buds or animals, which may then be killed by the return of winter. For several years I have often seen buds of post oak trees (*Quercus stellata*) in Oklahoma open in winter and then die. Later, when temperatures become and remain warmer, another set of buds open in place of the damaged buds. Whether this loss of some of the buds may have long-term effects on the growth of the oaks is not known.

Global warming has occurred in the past, most notably at the end of the most recent ice age. At that time, animals simply migrated northward (and plants, too, as their seeds blew in the wind or were carried by animals). Today, this is no longer possible. Temperatures are increasing so rapidly that migration, at least for plants, may not be able to keep pace. At the end of the previous ice age, the North American and European continents had wide open spaces for plant and animal migration. But today, cities, highways, and farms block their movement. We have set aside "nature preserves" for wild plants and animals, but as the climate becomes warmer, they will be trapped in preserves no longer suitable for them. Ecologists are beginning to design preserves that are connected by corridors, which will allow plants and animals to move northward from one preserve to another in response to global warming. But such designs are still far from widespread implementation and will probably prove to be too little too late for the wild species.

Shifts in the geographical ranges of plants and animals, and other biological changes, may have direct impact on humans. Studies led by Fakhri Bazzaz of suggest that global climate change may enhance the growth, and the pollen production, of ragweeds.[30] Warm weather causes poison ivy to produce more irritant. And tropical mosquitoes, which spread numerous diseases such as malaria and dengue fever, may spread northward and reintroduce these diseases, which Americans have been privileged to ignore, to the United States.

There is some evidence that global warming may have already begun to induce evolutionary changes. Researchers have observed genetic alteration within species. Genetic changes have occurred in fruit flies similar to those that occur in hot and dry conditions.[31] Researchers have also observed what might be the beginning of the evolution of one species into two or more, a process called speciation,[32] as a result of global warming. In the European blackcap example mentioned above, the birds that overwinter in England mate mostly with one another rather than with the birds that overwinter in Spain and Portugal; that is, they are reproductively isolated from the birds that migrate from Spain and Portugal. This may eventually allow one bird species to diverge into two. I have also observed, but not investigated, what may be the start of speciation in a wildflower species, *Oenothera rhombipetala*, in Oklahoma. Most plants in this species live for at least two years. The first year they produce a cluster of leaves, and the second year they producing a flowering stalk. But I observed that a few plants produced flowering stalks during their first year of growth in 2007. They could not do so until November, but the warm weather permitted the flowers to open and pollinators (moths and native bees) to visit them. The plants that produce flowers during their first year do not cross-breed with the others.

These evolutionary changes are not necessarily bad. In fact, they should allow species to adjust to the changes associated with global warming—except that the changes are occurring too fast for evolution to help them. They illustrate the very broad impact of global warming on the natural world.

Inevitable Surprises

If global climate change occurred slowly and gradually, perhaps we would be able to adjust to it. But the clearest lesson that we can learn from the time since the previous glaciation is that global climate change has not been uniformly slow. The most striking example is the Younger Dryas cold snap already mentioned, in which a temporary reversal of the postglacial warming trend began suddenly, lasted a few hundred years, then ended even more suddenly. Warm temperatures returned within a decade. The climatic changes predicted by scientists may take place gradually over a couple of centuries, or they may occur within one generation. Because the processes by which these changes occur are imperfectly understood, surprises are inevitable.

With more sunlight energy being trapped in the climate systems of the earth, the result will almost certainly not be a uniform warming of the planet. Instead, scientists expect that climate fluctuations will become more extreme. In effect, global warming is stirring the climate pot more vigorously. The exact results will be impossible to predict. In the summer of 2007, for example, Arctic sea ice melted at a much greater rate than had been predicted. The summer Arctic ice minimum had been declining at 8.6 percent per decade until 2005, then it declined almost 20 percent in 2007. It is not clear whether this represents a long-term trend.[33] This sudden melting of ice was made famous when the long-fabled Northwest Passage, an ocean route across the top of North America, opened up for the first time in recorded history and when Lewis Pugh, a British endurance swimmer, jumped into the ocean at the North Pole.

Although we can predict that coastal areas will experience an increase in rainfall, we cannot predict the pattern of that rainfall. Stronger hurricanes, for example, will bring the abundant rainfall further inland. An experiment in a California grassland indicates that increased precipitation during the normally rainy winter would have little effect on the growth of plants or the dominant species of plants and invertebrate animals, but rainfall that extended the spring rains into summer would have a profound effect. What that effect would be is also unclear. In this experiment, there was an initial burst of growth in some plant species, followed by a decline in both plant growth and number of plant and invertebrate species in the later years of the experiment.[34]

One reason that the detailed patterns of climate change are unpredictable is that they occur by means of "tipping points" or "thresholds."[35]

A complex system may be in a relatively stable state for long periods of time, then suddenly change to another stable state when a threshold is reached. Historical and sociological examples abound, such as revolutions and the Reformation. Engineering examples are also well known, in which a building or bridge suddenly collapses. Everyone knows about the proverbial "straw that broke the camel's back." Many scientific processes occur by means of threshold events. Rocks move along geological faults, not gradually, but suddenly: a major earthquake may occur after a buildup of pressure crosses a threshold. Chemical reactions may not occur at all until threshold conditions are reached.

Climate change may occur slowly until some threshold incident catalyzes a cascade of events that feed on one another. Here are some examples. The Arctic ice cap reflects a great deal of sunlight back into outer space; this reflected light, called *albedo*, does not cause the earth to become warmer. But as the Arctic ice melts, the ocean that the ice formerly covered now absorbs more sunlight, and retains the heat. Global warming causes the ice to melt, but the melting of the ice also causes more global warming. Another example is that decomposition of organic matter in peat bogs and soils around the world, especially in the far north, releases greenhouse gases such as carbon dioxide and methane (see next section) into the atmosphere. Some of the "mammoth steppe soils" in Siberia have organic deposits more than 150 feet thick. Global warming causes more decomposition, and decomposition causes more global warming in the cold peat bogs; some bog lakes in Siberia no longer freeze in the winter because of the warmth created by decomposition. In tropical areas such as Indonesia, global warming dries out peat bogs and encourages them to catch fire. Global warming also causes drier conditions in the interiors of continents. These drier conditions allow more wildfires to occur—which is what happened in thousands of square miles of tropical forests in 2005. The fires put greenhouse gases into the air and, at least temporarily, destroy the forests that could remove the greenhouse gases from the air. The exposed soils become warmer, which causes them to release more carbon dioxide. Therefore, greenhouse gases cause global warming that causes the release of more greenhouse gases, resulting in more global warming. Finally, consider the glaciers whose movement toward the ocean has been accelerated by meltwater, as described previously.

The inherent unpredictability of climate change makes it difficult to prepare for the future. If we knew exactly when and how much the sea level will rise, we might be able to have orderly migrations and plan ahead for building seawalls. We could prepare for explosive hurricane seasons if we could predict them. It is always more expensive to react to an emergency than to prepare for gradual changes. This is one of the reasons that many governments, especially in Europe, prefer to minimize global climate change than to deal with it. Insurance companies, and the reinsurers that insure them, also prefer to avoid paying for catastrophes.[36]

Perhaps the greatest impact of climate unpredictability is on national security. A report, to which retired American generals Anthony Zinni, Charles F. Wald, and Gordon Sullivan contributed, concluded that global warming is a national security threat. Many parts of the world are already in or on the verge of conflict. Global climate change could cause food production in some regions to collapse, water supplies to be interrupted, sea levels to wipe out coastal areas, and diseases to spread. The result would be millions of environmental refugees flooding across national borders. International conflicts are nearly inevitable. Because of international terrorism, and because American military interests are worldwide, the United States would inevitably be drawn into these conflicts.[37] The U.S. Department of Defense had placed an earlier memorandum, making similar predictions, on its Web site, but senior Bush administration officials removed it. The issue will not go away, however. Droughts, floods, disease, and the spread of insect pests were mentioned as likely consequences of global warming in an April 2007 report by the Intergovernmental Panel on Climate Change (IPCC).[38] In 2007, the World Health Organization announced that global warming threatened the livelihoods and health of millions of people, particularly in those countries that put the least carbon dioxide into the air. On June 24, 2008, the National Intelligence Council, which coordinates the work of all sixteen U.S. federal intelligence agencies, presented a report to Congress that indicated that global climate change may cause national security problems for the U.S. by destabilizing many countries around the world. This is the first official report from the U.S. government that links global warming to national security.

But even the most sophisticated computer models of what will happen in the future can be erroneous. Leading climate scientists announced in 2007 that this is already proving to be the case. In an earlier 2001 report, the IPCC predicted future increases in atmospheric carbon dioxide, global temperature, and sea level. With each prediction, the IPCC scientists specified an error range, which means the increases might be less, or might be greater, than they had predicted. Since 2001, all three factors have been increasing even more rapidly than the IPCC predicted—and the increases have been *right at the very top of the range of predicted possibilities*. The climate predictions, as frightening as they have been, have already proven to be underestimates.[39]

Atmospheric Carbon Dioxide Is a Cause of Global Warming

How Greenhouse Gases Contribute to Global Warming

The "greenhouse effect" makes the earth warmer than it would otherwise be. In a greenhouse, sunlight shines in through the glass roof and heats up the plants and benches, which conduct heat to the air. The glass roof holds in a lot of the heat from this warm air and therefore acts as a largely one-way portal for heat to enter the greenhouse. "Greenhouse gases" perform an analogous service for the earth's atmosphere. Light shines through the atmosphere and warms the earth's surface, which then emits infrared radiation—low-energy wavelengths of light that are invisible to us. The earth's atmosphere consists mostly of nitrogen and oxygen gases. In an atmosphere that consisted only of nitrogen and oxygen gases, all of the infrared radiation would be lost into outer space. Greenhouse gas molecules, however, absorb the infrared radiation, which makes them warm. They share their warmth with the other gases in the atmosphere, making the whole atmosphere warmer. The more greenhouse gas molecules there are in the atmosphere, the warmer that atmosphere, and the planet, will be.

Carbon dioxide is the most famous greenhouse gas. The atmosphere contains only about 380 parts per million (ppm), or 0.038 percent, carbon dioxide by volume—fewer than four molecules out of every ten thousand in the atmosphere. But without this carbon dioxide, the earth would be uninhabitably cold. There are other greenhouse gases in the atmosphere, including water vapor and methane (the major component of natural gas). Methane is about a thousand times less common than carbon dioxide in the atmosphere, but is at least twenty times as potent at retaining heat. However, of these three gases, only carbon dioxide can accumulate indefinitely. Water vapor in the lower atmosphere can build up until it condenses as clouds and falls as rain. Methane reacts with oxygen gas, producing carbon dioxide. Other greenhouse gases, such as nitrogen oxides and ozone, are even more potent and (in the case of ozone) persistent, but are rare compared to carbon dioxide. Even though methane is rare and persists only about a decade in the atmosphere, scientists estimate that it is responsible for about 40 percent of the global temperature increase.[40]

The greenhouse effect is, to a certain extent, a good thing. Without it, the earth would be about as cold as Mars now is. Mars itself has an atmosphere consisting mostly of carbon dioxide, but its greenhouse effect is minor because its atmosphere is so thin. The atmosphere of Venus contains huge amounts of carbon dioxide, which is the reason that Venus is hotter than Mercury even though it is further from the sun. But it is possible to have too much of a good thing. Too much carbon dioxide can cause the earth to become too warm for the life forms that are now living on it. This is a major reason that scientists view the increasing concentration of greenhouse gases in the earth's atmosphere with some alarm.

Atmospheric Carbon Dioxide Levels Are Increasing

It certainly wasn't what I was expecting to see in Hawaii. I stood near the summit of Mauna Loa, one of two large volcanoes that make up the Big Island. Everywhere I looked, I saw barren lava. It was cold and dry, with no plants in sight. I was part of a 1992 tour group of scientists who visited the U.S. National Oceanic and Atmospheric Administration (NOAA) observatory near the top of Mauna Loa. It is a meteorological, not an astronomical, observatory, and its purpose is to study the atmosphere. The top of a volcanic mountain in the middle of the Pacific Ocean is the ideal place to do this. The prevailing winds come eastward across the Pacific, over thousands of miles of largely uninhabited water, bringing air that is as close as you can get to an average sample of the global atmosphere. In 1958, the late atmospheric scientist Charles Keeling of the Scripps Institute of Oceanography began measuring carbon dioxide concentration in those atmospheric samples. The measurements continue today, a half-century later (fig. 3.6).[41] Keeling and colleagues determined the carbon dioxide concentrations with a spectrophotometer, which measures the amount of infrared radiation that an air sample absorbs. During my 1992 visit, I saw the original

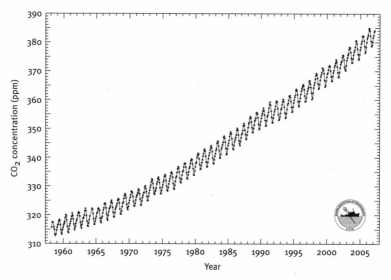

FIGURE 3.6. Carbon dioxide concentrations in air samples taken at Mauna Loa in Hawaii from 1958 to 2007. These samples are as close as we can get to global averages of atmospheric carbon dioxide levels. Each year, carbon dioxide levels have increased, and each year there is a fluctuation that largely represents photosynthesis of land plants in the northern hemisphere. Figure courtesy of Scripps Institute of Oceanography, University of California at San Diego.

spectrophotometer that was first used, and was still in use, although it was becoming difficult to find new vacuum tubes to replace old ones.

Two facts became obvious to Keeling after the first few years of measurement, and they have been confirmed every year since the measurements began. First, atmospheric carbon dioxide levels are increasing. The atmosphere contains only about 380 ppm of carbon dioxide. But each year, the concentration has increased by a little more than 1 ppm. This does not sound like much, but when Keeling began collecting data, the atmosphere contained only about 316 ppm of carbon dioxide. Scientists estimate that before the Industrial Revolution, it contained only 280 ppm. Carbon dioxide levels have increased 20 percent just since the measurements began, and 36 percent since before widespread industrialization. Therefore atmospheric carbon dioxide levels have been rapidly increasing for the past half-century, and probably for the past two centuries. At current rates, the atmosphere will have 500 ppm of carbon dioxide by 2050, and 750 ppm by 2100.

Second, atmospheric carbon dioxide levels fluctuate by about 5 ppm each year, increasing in the northern hemisphere winter, decreasing in the northern hemisphere summer. This is overwhelmingly the effect of land plants in the northern hemisphere, such as those in forests and grasslands. Plants remove carbon dioxide from the air through the process of photosynthesis (chapter 2). Some aquatic microbes do this also, but land plants show a strong seasonal pattern: a lot of photosynthesis during the summer, and much less during the winter. Aquatic microbes and tropical forests remove carbon dioxide more or less uniformly all year, unlike temperate forests. It is the northern hemisphere summer that is important in the annual decline in carbon dioxide, because the southern hemisphere is mostly water. Five ppm, therefore, represents the approximate amount of carbon dioxide that temperate land plants can remove from the global atmosphere in a growing season.

Both carbon dioxide and temperature have been increasing during the past half-century. This strongly implicates carbon dioxide as a major cause of global warming. But to demonstrate that the parallel changes are no mere coincidence, it is necessary to look further back into the past. In order to estimate global temperatures of the past, scientists have had to use proxy methods, as explained in a previous section. One of these methods is oxygen isotopes in the layers of ice from Greenland and Antarctica. But for atmospheric carbon dioxide, it is possible to make direct measurements instead of relying on proxies. As already mentioned, within each layer of ice, bubbles of air have been trapped, many visible to the unaided eye. The atmospheric gas contained in a layer of ice can be released in a confined laboratory container, and its carbon dioxide concentration can be measured. Therefore, scientists can make a temperature estimate and a carbon dioxide measurement for each layer of ice for the past several hundred thousand years.

The graphs of global temperature and carbon dioxide line up almost exactly (fig. 3.5).[42] Whenever the air had a lot of carbon dioxide in it, the temperature was greater than at times when there was less carbon dioxide in the atmosphere. This is a striking, and convincing, proof of the correlation of carbon dioxide with global warming. Correlation, however, is not the same as causation. I have already explained how an increase in carbon dioxide concentration can cause global warming. But could global warming cause an increase in carbon dioxide? Yes. Warmer temperatures stimulate decomposition by bacteria and fungi in the soils of the world; decomposition releases carbon dioxide into the air. Increases in carbon dioxide, and global warming, therefore cause each other. This accounts for the correlation. In the past, ice ages came and went, largely because processes such as the rotation and revolution of the earth initiated cycles of warming and cooling. A warming trend would stimulate the release of carbon dioxide, which would accentuate the warming trend. Before civilization, global temperatures appear to have increased prior to the increase in carbon dioxide. Today, however, it is the other way around. Atmospheric carbon dioxide has increased to its highest level in almost a half million years, and global temperatures have yet to catch up. For this reason, even if we stopped putting carbon dioxide into the atmosphere right now, global warming would still occur. The difference between what happened before civilization, in which global warming caused increases in carbon dioxide, and what is happening today, in which carbon dioxide is causing global warming, is that billions of humans are producing vast amounts of this gas.

Increased Carbon Dioxide Levels Are Largely Responsible for Global Warming

Carbon dioxide is not the only factor that may contribute to global warming. Others include the natural climatic variability caused by the movement of the earth and changes in the intensity of sunlight. The advance and retreat of glaciers across North America and Europe appear to be caused by changes in the movements of the earth relative to the sun, which produce an approximately one hundred thousand-year cycle of cold (glaciation) and warm (interglacial) periods. However, the changes in solar intensity are very small. By themselves, they seem unable to explain the ice ages. Apparently these variations in solar intensity have stimulated a series of processes that, working together, have produced ice ages during the past two million years. Although the ice ages have followed the pattern imposed by cycles of solar illumination, recent global warming cannot be attributed to it (table 3.1).

Other human activities may counteract global warming, but at a cost. The destruction of forests, and their replacement by barren land, reflects more sunlight back into outer space but also reduces photosynthesis. Particulate air pollution (such as smoke and sulfur dioxide) reflects sunlight back into outer space but creates environmental and health problems. In fact, the reduction of particulate air pollution in recent decades, allowing more sunlight to reach the surface of the earth, may be the reason that global temperatures have risen sharply in recent decades, while atmospheric carbon dioxide has been steadily accumulating for a century and a half.[43] As table 3.1 shows, none of these other processes has anywhere near the effect that carbon dioxide has had on recent global warming.[44]

Human activities have increased the amount of energy at the earth's surface by 0.2 watts per square foot (table 3.1). This is the equivalent of having an additional heat load of two hundred 30-watt light bulbs burning continuously on each acre of the earth's land and ocean surface. This is approximately equal to the amount of heating that is attributable to the buildup of carbon dioxide in the air. Carbon dioxide and other greenhouse gases have also been a major influence in the earth's

TABLE 3.1. Effects of Various Factors on the earth's Surface Heat

FACTOR	DIFFERENCE IN HEATING (WATTS PER SQUARE METER)
Human-caused	
Carbon dioxide	+1.66
Other greenhouse gases	+0.98
Ozone	
In lower atmosphere	+0.35
In upper atmosphere	−0.05
Water (methane breakdown, upper atmosphere)	+0.07
Reflected light	
Destruction of natural plant cover	−0.20
Black carbon from pollution	+0.10
Air pollution particles	
Reflection of sunlight	−0.50
Enhanced cloud formation	−0.70
Contrails from jets	+0.01
Natural	
Changes in solar brightness	+0.12
Net human effect	**+1.60[a]**

SOURCE: Intergovernmental Panel on Climate Change, *Climate Change 2007: The Physical Science Basis. Summary for Policymakers*, available at http://www.ipcc.ch/SPM2feb07.pdf.

NOTE: 1 square meter = 10.5 square feet.
[a]Because of error estimates, this number cannot be obtained by simple addition of the components.

long-term history: during the past two million years,[45] during the climate change that occurred about 30 million years ago,[46] and during the sudden warming that occurred about 55 million years ago.[47] At only one time, about 300 million years ago, has a period of ice ages given way to a prolonged warm period, and this event was apparently caused by an increase in atmospheric carbon dioxide.[48]

What Is Causing the Increase in Carbon Dioxide?

Although in previous millennia the changes in temperature and carbon dioxide were related to the ice ages, the present-day increase in carbon dioxide is undoubtedly the result of human activities. The main sources of carbon dioxide emissions are transportation and manufacturing. The burning of fuels such as coal, oil, and natural gas for transportation and industry releases huge amounts of carbon dioxide into the air.[49] Since the beginning of the Industrial Revolution, humans have burned more and more of these fossil fuels (so-called because they were formed millions of years ago, primarily from dead plants and microbes) each year. Of these three fuels, coal (often used in power plants and factories) emits the most carbon dioxide per unit of energy produced, because it already contains so much carbon. Oil, including oil refined into gasoline, also produces a lot of carbon dioxide from the tailpipes of the billion cars and trucks in the world. Another large source of carbon emissions is the burning of firewood. Firewood produces about 30 percent of the carbon dioxide that enters the atmosphere. This is the principal (or only) source of energy for billions of poor people.

Another reason for the prodigious increase in carbon emissions is that we are destroying plants. The destruction of forests and other plant habitats puts carbon dioxide into the air in two major ways. First, when the trees are burned or decompose, they release carbon dioxide into the air. Second, dead plants can no longer carry out the photosynthetic process that removes it from the air. In many instances, the natural habitats do not recover. In such cases, the loss of plant cover represents a permanent release of carbon dioxide and reduction in the ability of plants to cleanse the air of excess carbon dioxide.

The sheer volume of these processes—all of them caused by humans—accounts for almost all of the increase in atmospheric carbon dioxide. It is ludicrous to assert that we cannot know why its level is rising. One U.S. Republican congressman, Dana Rohrabacher of California, speculated in 2007 that a global warming event 55 million years ago was caused by unknown and unknowable processes—his guess was that it might have been dinosaur flatulence.[50] This suggestion is as irrelevant as it is ridiculous. Even if the causes of global warming in the past were not fairly well understood, the cause today is not in doubt. Nearly all scientists agree that, as explained in this chapter, human activities are the main cause of the global warming that is now occurring.

The agreement of scientists is perhaps best indicated by the strong conclusions issued by the IPCC. In February 2007, the IPCC stated that global warming from carbon dioxide was a certainty, and that the human contribution to this warming was 90 percent certain.[51] In April, the panel released a second report that predicted droughts in some areas, flooding in others, and the spread of disease and insect pests as consequences of global warming.[52] In May, it released a third report that proposed possible solutions.[53] In November, it produced a synthesis report that summarized all of the organization's conclusions.[54] When this international panel of scientists issues a consensus statement, it is as close to a scientific fact as we can get about climate change. The panel's statements are reviewed by representatives of many governments and therefore represent something as close to world consensus as we can reasonably expect to have.

Atmospheric carbon dioxide levels have been much higher in the distant past than they are now. And global warming is nothing new on this planet. Before 25 million years ago, Antarctica was fully forested. Then carbon dioxide fell below 500 ppm, and ice sheets began to expand in Antarctica. No competent scientist claims that global warming or the accumulation of atmospheric carbon dioxide will prove directly lethal to life on this planet. In fact, the earth was greener when there was more carbon dioxide in the air millions of years ago. But the consequences of global warming, as outlined above, may be disastrous to the habitats and ecological relationships of plants and animals, as well as to the human economy and to global politics. Never before has the earth had to support six and a half billion humans. Our entire system of agriculture and commerce is based on the assumption that global climate patterns and ocean levels will remain as they are except for occasional droughts or storms. This comfortable belief will almost certainly be shaken. It will not be necessary for global warming to make us fall over dead in order to wreak havoc on our global civilization. Human activity is pushing temperatures and carbon dioxide levels outside of the range that they have occupied for the past two million to three million years, at which time continents and global ocean circulation assumed their present patterns. There is thus no comparable situation in the past to which we can look that will allow us to predict what will happen next. The economic consequences, from higher sea levels, from storms, from disease, and from agricultural losses, are unpredictable and will probably be massive.

Can Plants Save Us from the Greenhouse Effect?

For the future security of the human and natural world as we know it, it is essential that the greenhouse effect be brought under control. Some scientists have made wildly speculative proposals to do this by reflecting sunlight back into outer space by launching giant mirrors into orbit or by spewing massive amounts of sulfate particles into the upper atmosphere.[55] Most scientists are convinced that the greenhouse effect can be reduced only by managing the amount of greenhouse gases in the atmosphere. The two general ways to do this are to release fewer greenhouse gases into the atmosphere and to remove them from the atmosphere. Some scientists have proposed that we reduce carbon emissions through "carbon sequestration," or chemical removal of carbon dioxide from effluent gases before they enter the air—for example, by removing carbon dioxide from the smokestacks of power plants. The technology to do this on any meaningful scale is currently experimental and expensive. A few other scientists have championed the chemical removal of carbon dioxide from the air.[56] This method would be even more uncertain and expensive. Thus, most scientists say that we should produce less carbon dioxide in the first place and remove it from the air by enhancing natural processes such as plant photosynthesis.

Photosynthesis is not the only natural process that removes carbon dioxide from the air. Over a very long period of time, the weathering of some kinds of rocks removes carbon dioxide from the air. Carbon dioxide reacts with silicates in the earth's crust, forming carbonates (limestone) that are washed away and accumulate as sediments on the ocean floor. The cycle is completed when continental drift thrusts the sediments into the crust and the carbon is released as carbon dioxide from volcanoes. This process acts as a regulatory mechanism, because a greater abundance of atmospheric carbon dioxide produces a warmer climate that stimulates the reaction between carbon dioxide and silicate. This process cycles a much greater amount of carbon through the earth than does photosynthesis. However, it requires millions of years and cannot rescue the earth from an excessive greenhouse effect on any time scale meaningful to humans. Photosynthesis is the only natural process that can quickly remove carbon dioxide from the air. If we promote the growth of plants, we may not only stimulate photosynthesis but even the formation of limestone: plants speed up the process of silicate weathering by acidifying the soil with their leaf litter.

At least one "think tank" of American antienvironmentalists, the Competitive Enterprise Institute, has claimed that photosynthesis is the reason that we do not have to worry about global warming. We can pour all the carbon dioxide we want into the air, and plants will clean up the mess. Their advertisements asserted: "Carbon dioxide—we call it life!"[57] As a botanist, I find this scenario very appealing—plants will save the world!—but unfortunately this argument is lethally simplistic. Perhaps the most obvious fact that contradicts this hypothesis is that plants are not, right now, absorbing the excess carbon dioxide in the atmosphere. Also, at the very time that we need this help from plants, we are destroying them. Furthermore, plants carry out photosynthesis only if they have enough nutrients and water. If large continental areas experience droughts, the trees will not remove very much carbon from the air. Researchers at Duke University performed a free-air carbon enhancement (FACE) study in a forested area of North Carolina. In FACE studies, the air is enriched in carbon dioxide, but the plants are not enclosed by any kind of barrier. The researchers increased the carbon dioxide in forest plots by 50 percent over the course of an entire decade. They announced in the summer of 2007 that the extra carbon dioxide had aided tree growth only in those areas that had plenty of water and good soil. Other studies have found that global warming might enhance soil and plant respiration, which releases carbon dioxide, even more than it stimulates photosynthesis, which removes it.[58] Finally, there are just not enough plants in the world to counteract all of the carbon that human activities are putting into the air. We need more plants.

We not only need more plant cover in the world, but we need plants that remove carbon from the air permanently. As stated earlier, each northern summer, plants remove about 5 ppm of carbon dioxide from the air. Most of this is returned, however, through decomposition of leaf litter. This is particularly true of short-lived plants such as crops. Researchers at the University of Illinois used FACE experiments to enhance the growth of soybeans and observed only modest increases, far less than the increases that had been predicted from laboratory experiments.[59] Other recent experiments have shown that extra carbon dioxide in the air may boost the growth of forests, but that these results may be transient—forests do not continue to show year after year of additional growth in the presence of elevated carbon dioxide.[60] If, in contrast, forests are allowed to store carbon in the form of wood, the carbon has been removed from the air for the life of the tree. The trees that remove the most carbon from the air during their lives are also the ones that grow slowly. A huge California redwood may remove 30 tons of carbon from the air—but may require two millennia to do so. Preserving forests and planting more trees is a necessary but not entirely sufficient part of removing excess carbon dioxide from the atmosphere.

Another way in which plants may be able to remove carbon dioxide from the atmosphere for a long period is by the growth of roots. When dead roots decompose, much of the carbon goes into humus rather than directly into the atmosphere. So if we grow more plants, more carbon will be stored up in the soil. In soils all around the world, humus contains a massive amount of carbon. The cold, soggy soils of Siberia alone may hold 70 billion tons of carbon. However, global warming is causing soils everywhere to release carbon dioxide, which will enhance global warming still further.

But even if we plant as many forests and grasslands as possible, they will not remove all of the additional carbon we are pouring into the atmosphere. Plant ecologist Christopher Field and colleagues estimated that the earth's natural ecosystems produce 117 billion tons of biomass per year.[61] This biomass consists of either carbohydrates, or material derived from carbohydrates, which are produced by photosynthesis. Most of the mass of these carbohydrates comes from carbon

[76]

dioxide. The burning of fossil fuels produces about 7 billion tons of carbon, which corresponds to about 25 billion tons of carbon dioxide, each year. If plants and aquatic microbes were to absorb the entire 25 billion tons, it would be necessary to increase the growth of plants worldwide by about 21 percent relative to the plant cover that existed before civilization. We will probably never be able to restore the earth's vegetation cover to its premodern state (chapter 1), much less enhance it by 21 percent. Nevertheless, this figure suggests that it is not physically impossible for newly planted forests to cleanse the air of the carbon dioxide that we add to it.

Not all plant cover would be equally helpful in reducing the greenhouse effect. Boreal forests (of the far north) absorb much more sunlight than does snow cover. This causes the trees to become warm and release heat into the surrounding air, which would not happen with sun shining on snow. An expansion of boreal forests at the expense of tundra, which is covered by snow except for a couple of months each year, would therefore result in a net warming of the planet.[62] This point is true but may be moot, because with global warming, the snow cover of the tundra will be severely reduced anyway, and scientists predict that the boreal forest will move northward whether we plant trees or not. Moreover, it will do no good to plant trees under conditions in which they cannot survive. Millions of trees have been planted in arid, dusty regions of China, but many of these have died. The restoration of grasslands would have been a more effective way of stabilizing the soil and of removing carbon from the air.[63] Grasses have massive root systems and are particularly good at putting carbon into the soil.

We should not, however, underestimate the ability of plants to remove carbon dioxide from the air. At the time, about 450 million years ago, when plants began to live on land, atmospheric carbon dioxide concentration was fifteen times as great as it is now. Because carbon dioxide was so abundant, these plants could easily absorb it through their green stems. They removed carbon dioxide from the air by photosynthesis, as well as by stimulating chemical reactions in the soil that transformed it into calcium carbonate (limestone). But as they continued removing carbon dioxide from the air, it became scarce enough that their chubby stems could no longer absorb enough for their photosynthesis. Some of these plants evolved leaves, which are very thin photosynthetic structures, and continued removing carbon dioxide from the air. They did this so effectively that, according to plant ecologist David Beerling, they caused the opposite of a greenhouse effect—they caused glaciation to occur in the middle of the vast southern continent that was centered on the South Pole at that time.[64] If Beerling is correct, then plants can certainly alter the climate of the globe and do so in a way that helps reduce the greenhouse effect.

The oceans also absorb carbon dioxide. Many aquatic microbes carry out photosynthesis just as land plants do. Many small aquatic organisms, and some that are not so small, produce shells of calcium carbonate. When these organisms die, their shells, and the carbon they contain, settle to the bottom of the sea, forming organic limestone. In addition, ocean water itself absorbs carbon dioxide, producing carbonic acid. Currently, by these processes, the oceans remove about 2 billion tons of carbon from the atmosphere—about one-fourth of what is added—each year. We cannot, however, trust these processes to cleanse the air of excess carbon dioxide. Absorption of carbon by the oceans may not continue indefinitely at its current rate. Carbonic acid in the oceans may reduce the growth of the organisms that now make organic limestone, and warmer oceans will absorb less carbon dioxide.

Each person leaves an "ecological footprint" on the world. This refers to the acreage of productive land and water (such as forests and agricultural fields) that would be required to remove the carbon dioxide from the air that a person is responsible for releasing and to provide the raw materials for that person's activities on a continuing basis. This includes the absorption of the carbon dioxide not only from what each person directly releases, such as from an automobile tailpipe, but the carbon dioxide released by power plants and factories that provide the person with what he or she consumes. The average American has an ecological footprint of 25 acres, almost five times the world average, and much higher than other industrialized nations (table 3.2).[65] A "carbon footprint" is the amount of carbon dioxide that a person directly or indirectly releases into the air.

TABLE 3.2. Comparisons of the Ecological
Footprints (per Person) of Selected
Countries versus the World (2007)

COUNTRY	HECTARES[a]
United States	10
Canada	8
United Kingdom	6
Japan	4
Germany	2
China	2
World	2
India	1
Kenya	1

SOURCE: Global Footprint Network, "National
Footprints," available at http://www.footprintnetwork
.org/gfn_sub.php?content=national_footprints.

NOTE: 1 hectare = 2.5 acres. An ecological footprint is
the number of hectares of productive land (such as
forests and agricultural lands) necessary to supply the
resources used by an average citizen of the country.
[a] Rounded to the nearest hectare.

The average citizen of India has an ecological footprint more than ten times smaller than that of an average American and produces almost thirty times less carbon dioxide. If all six billion people in the world had the ecological footprint of an American, four worlds of plant growth would be required to compensate for it. The United States is also the leading producer of carbon emissions on a per capita basis (table 3.3).[66]

It is highly unlikely that any American would ever be able to plant enough trees to make himself or herself carbon neutral. One study estimates that it would require about thirty walnut trees to compensate for the carbon emitted by two family-size vehicles, and another thirty to compensate for the production of energy used in a typical American house on a continuing basis.[67] This estimate, however, is based on individual trees growing as rapidly as possible, rather than natural forests. My own calculations suggest that when carbon emissions are compared with the carbon uptake by natural forests, it appears that it may take up to sixteen acres of deciduous forest to counteract the carbon emissions of a single average American automobile.

Even though planting trees cannot, by itself, save the world from an excessive greenhouse effect, we can hardly expect success without saving

TABLE 3.3. Comparisons of Carbon Emissions from the Burning of Fossil Fuels in Selected Regions and Countries (2004)

REGIONS/COUNTRIES	CONTRIBUTION TO WORLD CARBON EMISSIONS (%)	METRIC TONS OF CARBON DIOXIDE PRODUCED PER PERSON
Africa	3.6	1.1
Australia	1.4	19.4
Central and South America	3.8	2.4
East and Southeast Asia	35.5	2.7
China	17.4	3.9
India	4.1	1.0
Japan	4.7	9.9
Eurasia	9.4	8.9
Russia	6.2	11.7
Europe	17.2	8.0
Middle East	4.9	7.2
North America	25.5	16.0
Canada	2.2	18.1
United States	21.8	20.2
World		**4.2**

SOURCES: Energy Information Administration, "Environment: Energy-Related Emissions Data and Environmental Analyses," available at http://www.eia.doe.gov/environment.html. Total emissions: http://www.eia.doe.gov/pub/international/iealf/tableh1co2.xls. Per capita emissions: http://www.eia.doe.gov/pub/international/iealf/tableh1cco2.xls.

NOTE: 1 metric ton = 1.1 English tons.

the trees we have, and planting more. Green plants are essential allies in the effort to prevent a disastrous greenhouse future. In the biblical story from which the epigraph to this chapter is taken, Jesus urged his listeners to look at the sky and draw conclusions beyond the present moment. We need to do the same today by looking at the temperature of, and carbon dioxide in, the air and drawing conclusions about the trees and about how we should live.

chapter four

SHADE
TREES MAKE GOOD AIR CONDITIONERS

In honor of your birthday, a tree has been planted in
Israel. Thursday is your day to water it.
—GREETING CARD

The time: about 1800.

The place: northwestern Georgia.

American soldiers are building a fort.

The soldiers are part of one of the largest transformations of the earth in human history. The young United States engaged in a centuries-long frenzy of destroying the natural landscape and replacing it with its ideal of an artificial one. Americans cut down nearly the entire eastern deciduous forest, replacing it with farms and cities. And they made deliberate, unconcealed attempts to exterminate the native inhabitants of this land, replacing the diversity of native cultures with a uniformity of language and economy. The soldiers building the fort in Georgia are deliberately working toward these ends: they cut down every tree as they build the fort, and they consider the Cherokees, in nearby villages, to be savages—uncivilized and unworthy to inhabit the land. Sweating in the hot Georgia sun, the soldiers seek shade under canvas and log. They sit and curse the heat, hanging out their tongues in anticipation of something to drink.

Nearby, Cherokee men, women, and children rest in the shade of the many large trees that are contained within the stockade fence that surrounds their village (fig. 4.1). A little creek runs through the settlement. They relax.

Both the soldiers and the Cherokees are in the shade, but only the Cherokees are cool in the hot Georgia summer. The reason is that they

FIGURE 4.1. Tsalagi, a reconstructed traditional village of the Cherokee tribe, in Tahlequah, Oklahoma. Notice that inside the walls of the village there are numerous trees. Photograph by the author.

repose underneath the cool, living shade of the trees, while the soldiers suffer under the dead shade of logs and fabric. The roots of the trees reach far down into the moist soil and draw water through the wood and into the leaves. The water evaporates from the leaves, cooling them. This process, *transpiration*, makes the tree into a gigantic green air conditioner. Of course, this gigantic green air conditioner does not actually make the heat energy disappear. Energy never just disappears. Rather, the energy is in the water vapor. But the water vapor, and the heat that it contains, rises into the air, away from the surface of the earth where plants, animals, and people live.

How Trees Use Water to Cool Themselves

Transpiration pulls water from the soil, through the roots and trunks, and into the leaves. Microscopic streams of water are pulled through tiny cells called *xylem vessels*, which I will here call "xylem pipes." There are xylem pipes in the roots, the trunk, the branches, and the leaves of the tree. Xylem pipes are the major component of wood, and are thus very abundant in the trunk. Xylem pipes look like plumbing pipes: they are long, narrow, and empty except for water. Because water molecules can stick together (a process called *cohesion*), the water molecules that evaporate into the air from the leaves literally pull the water molecules that are behind them, just like a rubber band being stretched. Also like the rubber band, the microscopic streams of water become narrower as they are stretched. When water is transpiring rapidly from the leaves of a tree, the trunk of the tree actually becomes measurably narrower.[1]

Not all of the xylem pipes have the same diameter. Some are relatively large (about one-twentieth of an inch across); others are much smaller. Xylem pipes with large diameters can conduct more water than narrow ones. And not just more, but disproportionately more. If a large pipe has a diameter ten times greater than the diameter of small a pipe, it can conduct far more water than the sum of ten of the smaller pipes. The reason is based in the science of fluid dynamics. In theory, one of the large pipes could conduct as much water as *ten thousand* of the smaller pipes. This mathematical pattern does not apply precisely to wood; xylem pipes are not perfectly round, and there are occasional constrictions in the flow of water up through the pipes. Still, lightweight wood with large-diameter pipes can conduct more water, and conduct it more rapidly, than heavy wood with small-diameter pipes.[2]

Sometimes, when the water is stretched too much, it can snap like a rubber band, a process known as *cavitation*. Sensitive microphones can actually hear the little snaps of water in xylem pipes in wood on a hot day. Although the big pipes can conduct more water, the water in these big pipes is more likely to snap than the water in the small pipes. Therefore, most kinds of wood consist of a mixture of big pipes (good for conducting water rapidly when it is abundant) and small pipes (which are good at avoiding cavitation when water is scarce).[3]

When the water gets to the leaves, it evaporates in the little air spaces inside the leaf. The water vapor then diffuses out into the air through little pores called *stomata* (Greek for "mouths"). Stomata can open and close: they allow rapid transpiration when open, and they stop transpiration when they close. The rest of the leaf surface is covered with wax, and water vapor cannot diffuse through the layer of wax.

A square inch of leaf surface cannot transpire as much water as would evaporate from a square inch of water surface, such as the surface of a lake. But trees have an immense amount of leaf area. The *leaf area index* measures the amount of leaf surface area relative to the ground area beneath it. Thus a leaf area index of 1.0 means that each square foot of ground is shaded by one square foot of leaves. Forests contain both canopy and understory trees. Although an individual tree may have a leaf area index of only 2.0 to 3.0, a forest can have a much greater leaf area index. According to a study that summarized leaf area index measurements that were made from 1932 to 2000, the coniferous forests of the far north have leaf area indexes of about 2.6 to 3.5.[4] Temperate forests have leaf area indexes of about 5.0 to 6.7. One oak-maple forest area in North Carolina had a leaf area index of 7.3.[5] Tropical forests have leaf area indexes of about 3.9 to 4.9. Tropical forests have a lower leaf area index because, although they produce more leaf area each year, they also shed their leaves faster. However, they maintain this leaf area all year, whereas temperate forests shed most of their leaves in autumn. In that forest in North Carolina, there was more than seven times as much leaf area as ground area—and all that leaf area was transpiring. Therefore, if you stand directly under a tree and look up, you are looking through an average of two to seven layers of leaves.

The rate of transpiration depends on many factors: there is more transpiration when the weather is relatively warm and dry, but less transpiration if a drought causes the stomata to partially close. Leaves down in deep shade may transpire much less than those out in the sun. Taken together, the immense amount of leaf area on a tree or in a forest can allow a truly astounding amount of transpiration. Estimates vary widely, but one reliable approximation is that a large broad-leaved tree can transpire more than 100 gallons of water each day during the growing season.[6] This could easily add up to almost 10,000 gallons of water over the course of a summer—from one tree. The earth's surface (excluding oceans) has 62 billion tons of plant growth each year,[7] and these plants are all putting water into the air. There is a truly prodigious amount of transpiration going on in the natural world.

Transpiration can remove a great deal of energy from the leaves. About 204,000 kilocalories of heat (referred to as "calories" in everyday conversation) are dispersed in the 100 gallons of water that a big tree transpires in a single day. This is roughly the amount of heat that would be released by burning about 80 pounds of biomass. Without transpiration, leaves quickly become hot and can be damaged or killed. Plants that have deep roots and access to abundant water in the soil have cooler leaves than plants, just inches away, that have shallow roots and less access to soil water.[8] Transpiration keeps the flow of water coming to the leaves. Photosynthesis and metabolism of leaf cells requires minerals such as calcium and potassium, which are dissolved in the water that transpiration pulls up to the leaves. But the main benefit of transpiration for the tree is that it provides evaporative cooling to the leaves.

Animals cool off by the evaporation of water also. Water is continually evaporating from and cooling your skin. Only when this process occurs rapidly do we notice the accumulation of sweat; but even when you are not sweating, you are experiencing evaporative cooling. For this reason, it is important to consume liquids, especially water, during hot dry weather, even if you are unaware of thirst. People with defective sweat glands frequently suffer heat buildup in their bodies. Trees are not transpiring in order to create cool shade for us, but when we are down in the shade, we can enjoy the benefits of these enormous green air conditioners.

Shade in the Human Economy

Americans today do the same thing that the soldiers at the Georgia fort did: when we decide to construct a new building, the first thing we do is to "improve" the land—we chop and bulldoze the trees. A whole tract of forest is being razed and burned outside my office window as I make final revisions on this manuscript. Chopping down the trees produces a flat, clear field on which to construct our dreams, and from which the world of nature, which still frightens us with its power and mystery, has been driven away. As a result, we build our houses and offices out in the brutal intensity of sunlight, where they get hot, and in which we swelter with discomfort.

Or, at least, we used to. Just a few decades ago, our ancestors lived and worked in buildings that baked in the summer sun. But today, we will not tolerate this discomfort. We insist on continual air conditioning, where we live, work, shop, and worship. All of our buildings, while quiet and cool inside, are noisy outside with the buzz of loud fans and compressors, like ice cubes covered with angry wasps. We consider air conditioning a right, and label negligent those civic officials who fail to make air conditioning available to poor people. The electric utility companies measure the amount of electricity that we use for air conditioning and for other purposes. They may refer to this electricity as not "use" or "consumption," but rather as "demand."

We still have not learned the lesson of the Cherokees and soldiers in the forest. Instead of letting the trees do the air conditioning for us, we kill the trees, then install large, noisy air conditioners that gluttonize our electricity. Air conditioning is one of the largest uses to which summer electrical generation is put; and it is frequently the cause of electrical power outages, as consumer "demand" for energy outstrips our ability to produce it.

Urban areas, consisting largely of buildings, streets, and parking lots, have very little plant cover. With little transpiration to cool them, many cities become "heat islands" that are much warmer than the surrounding forested areas. Studies in cities all over the world have confirmed this pattern. The National Aeronautics and Space Administration (NASA) has closely studied several cities (such as Atlanta, Dallas, San Antonio, Nashville, and New York), using temperature measurements from satellites. Day or night, urban Atlanta is about 6°C (10°F) warmer than surrounding suburban areas, and New York City (not including Central Park) can be more than 7°C (almost 13°F) warmer than suburban areas.[9] In urban areas, not only does all the sunlight reach the buildings, the ground, and the people, but dark surfaces such as asphalt absorb the sunlight and make the air warmer. It is not just your imagination that old, shady neighborhoods are cooler than new, treeless suburbs or inner cities. Meanwhile, as populations of cities continue to grow, more forests are cleared. Each decade, the Atlanta vicinity has lost about 10 percent of its forest cover. This makes the heat island effect progressively worse each year.

High temperature is not the only effect of urban heat islands. Although the ozone layer high in the atmosphere protects us from ultraviolet radiation, ozone near the ground is a dangerous pollutant. The warmer air of a city can encourage the formation of ozone smog. Moreover, urban areas are not only hot in the daytime but hold on to their heat at night. Hot air is lighter than cool air, so the cool air rushes into the city as the hot air rises, especially at night. This can produce rain, including an increasing number of heavy thunderstorms. This long-suspected connection between urban heat and increased rainfall was confirmed by satellite observations in 2002. Regions immediately downwind from cities can have almost twice as much rain as regions immediately upwind of cities. When wind blows across cities, the tall buildings create a rough surface that slows down the wind and encourages the air to rise, which also contributes to the higher rainfall over and downwind from cities. Data, compiled in 2006 from ground-based weather stations, has confirmed that a similar increase in rainfall has occurred in desert cities such as Phoenix, Arizona, and Riyadh, Saudi Arabia. Cities, therefore, create their own weather—by forming heat islands and generating some of their own rainfall.[10]

What can we do about the higher temperatures around buildings and in cities? The answer may be very simple, according to Dale Quattrochi, climatologist at NASA: just plant more trees. Trees not only create cool shade but also remove carbon dioxide from the air (chapter 3). New trees can help a city reduce the heat effect and compensate for the additional carbon dioxide and ozone.[11] Research at Lawrence Berkeley Laboratory has indicated that, in warm climates during the summer, a house shaded by trees on the south and west sides may require one-third less air conditioning as an unshaded house in order to maintain the same temperature.[12] Buildings are the single largest consumer of energy in the United States and are responsible for 38 percent of the carbon emissions in the country, much of it for air conditioning.[13] So shading buildings with trees can have a dramatic impact on our use of energy.

Planting trees is not always the solution for a city. Desert cities face water shortages, and trees, as noted above, transpire a lot of water. Some cities, like Phoenix and Santa Fe, have *xeriscape* instead of lawns in many new subdivisions in order to save water. Xeriscape uses rocks, along with plants such as cacti and bushes that are adapted to xeric, or dry, conditions. These places could not afford the water that transpiration would require. In cities with moderately dry conditions, it may be advantageous to plant trees, such as oaks, that transpire less water than trees such as maples. Also, some trees are stronger than others. Although a strong oak tree can withstand most storms, flimsy trees such as hackberry and silver maple frequently have rotten, hollow interiors, and may fall and damage a roof during a storm (as my family and I have discovered twice and at great expense in recent years). An intelligent choice of tree species is crucial. One tree that should never be planted to counteract the heat island effect is the tamarisk, a small tree from the Middle East. It does not cast very much shade, and it transpires a lot of water. It is, furthermore, a highly aggressive invader along desert rivers and seasonal streams. Planting trees is not always the solution for individual buildings, either. Many new buildings are not only taller than newly planted tree saplings but taller than the mature trees will ever be.

Even better than planting trees is to not cut them down in the first place. When constructing a new building, a contractor can choose to put the building underneath large trees, rather than cutting the trees down. Careless operation of construction equipment can damage tree roots even when the trees are not cut down, but this need not happen if prudent building techniques are used. Such careful, "green" construction might cost a little bit more money up front, but can save the investor—and the earth—energy and carbon dioxide costs further down the road.

Many cities have decided that the preservation of existing trees, and planting new trees, is a good investment. New York City Parks Commissioner Adrian Benepe, using a computer program developed at the University of California, Davis, estimates that the 592,130 trees in

New York public areas generate $5.60 in savings for each $1.00 spent to plant and maintain them.[14] A study of the urban trees of New Berlin, Wisconsin, estimated that their shade saved residents $11,000 in air conditioning costs, and this was expected to rise to $107,000 as the trees grow. The trees remove more than $478,000 worth of pollutants (primarily ozone) from the air of that city each year.[15]

Can anything be done to reduce the heat of the buildings that are too large to be shaded by trees? Yes. Many new buildings are constructed with a "green roof," and many old buildings have had green roofs added. A green roof is one that has plants growing, and transpiring, on it. Green roofs have a long history; Scandinavian villages have had sod roofs for hundreds of years. A roof without plants can reach very high temperatures in the summer sunlight; transpiration of plants can greatly reduce the heat load on the roof itself, much of which would otherwise have to be handled by the building air conditioning system.

An *extensive* green roof is one that is minimally reinforced and has a small amount of soil and little plants. Some extensive green roofs are little more than turf. An *intensive* green roof requires additional reinforcement to hold the weight of deeper soil and larger plants such as trees. Therefore, whereas an extensive green roof can be added to an existing building, an intensive one must usually be planned as part of the original construction. For example, intensive green roofs require careful insulation to prevent dampness from causing long-term damage to the ceiling inside. Modern building techniques, however, can handle this problem.[16]

The plants on the roof need water, just as do plants in the ground. Although rain can provide much of this moisture, intensive green roofs may require additional water. It need not come from limited municipal water supplies, however. Rainwater can be collected in cisterns for later use, and the plants can be irrigated with "gray water"—water that is not potable but is not sewer water either—left over from sinks in the building.

Green roofs have been used by private businesses and by governments worldwide. Many cities, like Chicago, have regulations designed to reduce the heat island effect, and green roofs are an increasingly popular way to meet those regulations. The Greenroof Projects Database lists 609 projects covering more than 444 million square feet, many of them in the United States.[17] The following are just a few examples. Germany, where about 10 percent of the roofs are "green," may lead the world in using this technology. The municipal government of Tokyo mandated in 2001 that any structure with a roof area of more than 10,000 square feet have green roofs on at least 20 percent of the area. The U.S. Library of Congress is covered by 100,000 square feet of green roof. The city hall of Chicago has an extensive green roof, which officials estimate saves $10,000 a year in building energy costs. Ford Motor Company's assembly plant in Dearborn, Michigan, has more than 450,000 square feet of green roof.

A green roof can offer benefits beyond just the reduction of heat load. A well-constructed green roof can last longer than a conventional one, largely because the plants buffer the roof from the effects of wind and erosion. Leaves soften the rainfall and allow water to percolate into the soil on green roofs just as in natural forests (chapter 5), and this can reduce the intensity of storm water drainage from the roof. In municipalities where sewage is mixed with storm water, a surge of storm water can overwhelm the capacity of a treatment plant to handle sewage, and untreated water overflows into rivers. The "sponge" effect of vegetation, including the vegetation on green roofs, can temporarily hold back some of this water. Moreover, a green roof on a lower building can be more visually pleasing to residents of apartments or occupants of offices in taller buildings nearby, increasing the price that can be asked for occupancy of those apartments or offices. Increasingly, green roofs are being seen as a sign of a progressive, livable, and attractive city. In urban areas of the third world, where residents may have limited access to fresh food, rooftop gardens can provide fresh and healthy produce. Green roofs, then, can be part of a package of sustainable development that will help to raise people around the world out of poverty.

It isn't hot all the time. What about winter? During winter, also, trees help buildings maintain a favorable temperature. In summer, the trees around our buildings shade us; in the autumn, they graciously lose their leaves and allow the sunshine to stream in and warm them up. Not only that, but their branches form a sweater around our buildings, even if a somewhat sparse one, that slows down the wind and insulates our buildings. Green roofs can enhance the insulation of the roof, helping the building to retain more heat in the winter. A study in Toronto showed that a green roof can reduce summer cooling demands by 25 percent and winter heating demands by 26 percent.[18]

A city with trees is cooler, which makes it a nicer place to live. But a city with trees is not only nicer because of cool shade but because of the restfulness of the trees themselves. Psychological studies have repeatedly shown that humans prefer to live in landscapes that have some mixture of forests and open areas, not unlike the savannas in which our species evolved. It is well known that children with attention deficit hyperactivity disorder (ADHD) have reduced symptoms if they have time for recess. Research by Frances E. Kuo and others at the University of Illinois has shown that ADHD symptoms were reduced more by outdoor play time than by indoor play; and by outdoor play in the vicinity of trees and other plants than by play on cement or asphalt.[19] In another experiment, volunteers watched a stressful movie, followed by pictures of either urban or natural settings. Those who watched the natural images recovered from the stress more quickly, as measured by blood pressure, heartbeat rate, and facial muscle tension. Yet other studies have demonstrated that patients experience less stress prior to surgery if the waiting room has plants, aquaria, or windows with a natural view. After surgery, patients recover more quickly and require less pain killer if they have a window with a view of trees. Studies with psychiatric patients and prisoners also show that views of nature can reduce symptoms of stress.[20] Cool shade makes us happier as well as more comfortable. Trees absorb noise as well, thus creating a quieter environment. Like many others, I try to keep a few layers of trees between me and the rushing mass of humankind.

However, this solution may ultimately fail. If we plant trees that grow slowly and strong, we must wait decades for their shade. But by the time they become large enough, global warming will have changed the climate enough to kill them (chapter 3). This is another reason that we should preserve existing trees rather than just planting new ones: the existing trees can shade us during their few remaining decades.

During one of his campaigns, the Persian leader Xerxes encountered a large, famous sycamore tree and marveled at the depth and extent of its shade. He appointed a soldier to guard the tree from harm, and his appreciation of its shade has been immortalized in the famous "Largo" of Handel's opera, *Serse* [Xerxes]. The biblical prophet Jonah lamented inconsolably when he lost his only comfort, the shade of a "vine" that many scholars believe was actually a castor bean bush. In our artificial world of air conditioning, we have begun to ignore the beauty and value of shade. For our survival, we will have to recultivate the appreciation of shade.

chapter five

THE WATER CYCLE
PLANTS PREVENT DROUGHTS AND FLOODS

And thou beholdest the earth blackened; then, when
We send down water upon it, it quivers, and swells, and
puts forth herbs of every joyous kind.
—KORAN 22:5

The Water Comes Down

Clouds gather ominously in the sky. A thick blanket of water vapor precipitates into torrents of rain. Cities and villages flood, washing away lives and livelihoods. The story is repeated over and over, from Johnstown, Pennsylvania, in 1889 to Honduras in 1998. People call them acts of God, as if the Almighty is responsible for causing these catastrophes. Clergy and laity unite to pray for God to bring relief to the suffering. But perhaps God had already answered their prayers—beforehand. In each of these places, the hillsides had once been clothed with trees, and those forests had helped to protect human towns and cities from floods and mudslides. But the people had cut many of them down. Often, reporters overlook the fact that part of the blame rests on the destruction of the trees. For example, a news report in the late 1990s described mudslides in Nepal without ever mentioning the local deforestation, which could be easily seen in the photograph that accompanied the report.

Forests and shrublands grow profusely on mountain slopes around the world—not just the mountains with good soil and abundant rain, like the rich forests of the Appalachians and Great Smoky Mountains in eastern North America, but even in very poor and dry soils, like the White Mountains of California and Nevada. The forests and shrublands are the protectors of the entire watershed. To visualize the many ways in which forests protect the mountains and the valleys below them, let us trace the pathways of raindrops as they fall from the sky. Some of the drops collide, at full speed, with the foliage of the trees. When, at last, they drip down to the soil, their speed is much reduced. The momentum of the raindrops is further slowed by understory wildflowers and leaf litter. By the time the drops reach the ground, they form a slow trickle, which penetrates between the soil particles and downward into deeper layers. Other raindrops drain down the branches and trunk, and percolate slowly into the ground. The roots of the trees and understory plants hold the soil, thus the small amount of water that runs off from the forest contains very little mud. When the rain stops and the dry season begins, the forest trees are able to draw on the water that has penetrated and been stored in the soil. It is not only the trees, however, that benefit from this underground water. Some of it penetrates deeply enough to replenish wells, especially those downhill from the forest. It is the *groundwater* that keeps the creeks and rivers running even during rainless seasons.

In contrast, when raindrops collide with bare ground at full speed, they dislodge soil particles and carry them down the surface of the hill in little rivulets of not just water but also of mud. These rivulets converge into raging brown torrents and merge their contents to make muddy rivers overflow their banks. As a result, a heavy rain causes floods and mudslides from bare hillsides, instead of penetrating into the ground and replenishing underground stores. Therefore, in the dry season, the rivers have less water flow and the wells less water, if the hillside above them is barren (fig. 5.1). Half of the rain falling on bare soil runs off, in contrast to about one-fourth of the rain falling on farmland, and less than 1 percent of the rain that falls on grasslands and woodlands (table 5.1).[1] We have probably all seen something similar to what I am watching from my office window as I write this manuscript: a large tract of forest has been cut down, and a heavy rain is rushing from this clearing onto the road below—and taking a lot of mud with it.

Throughout human history, the destruction of forests has been followed by floods. The hills above Johnstown, Pennsylvania, had been deforested, just like most of the rest of the oak and maple forests of the

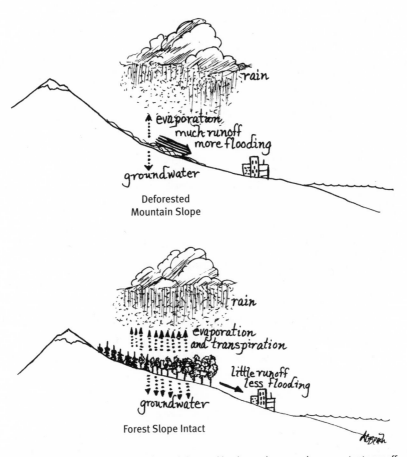

FIGURE 5.1. These drawings show that a deforested landscape loses much more water to runoff and flooding than does a landscape with intact forests, and that intact landscapes contribute water vapor to the air through transpiration. Illustration by Gleny Beach.

eastern United States. Heavy rains overwhelmed the feeble flood control devices that had been constructed. The tropical forests of Honduras have been devastated, as they have been in almost every tropical country. When it struck the coast of Honduras in 1998, Hurricane Mitch poured down almost a year's worth of rain in two hours. No forest could have prevented the resulting floods, but the damage was worse than would have occurred if the forests had been intact. Haiti and the Dominican Republic

TABLE 5.1. Amount of Soil Erosion and Water Runoff from Soils with Different Plant Cover.

PLANT COVER	SOIL LOST (METRIC TONS PER HECTARE PER YEAR)	PERCENTAGE OF RAINFALL LOST AS RUNOFF
Ungrazed thicket	0	0.4
Grass	0	1.9
Millet	70	26.0
Bare ground	146	50.4

SOURCE: Based on data from Charles L. Redman, *Human Impact on Ancient Environments* (Tucson: University of Arizona Press, 1999), 101.

NOTE: 1 metric ton = 1.1 English tons; 1 hectare = 2.5 acres.

share the island of Hispaniola in the Caribbean. Haiti has suffered a much greater amount of deforestation than the Dominican Republic (fig. 5.2). Although both countries experience periodic hurricanes and floods, Haiti consistently suffers more damage from them.

Every few years in California, heavy rain bombards the coastal mountains and washes away homes. If heavy rainfall occurs in the winter following a chaparral fire or after an oak woodland has been cut down, floods and mudslides can result. The shrubs and forests would have held back much of the flood water. There is no way that the floods in the burned chaparral could have been prevented, because fire is a natural and inevitable part of these shrublands, whether people live there or not (chapter 9). But the flood intensity in areas once covered with live oak woodlands could have been reduced if the oaks had been allowed to remain. Even in the chaparral areas, most of the danger of flooding is in the year following a fire. After a fire, the chaparral bushes resprout, protecting the watershed. But if the oaks are cut down, the watershed is left unprotected.

The above story presents an accurate general picture. However, as is always the case with ecological processes, the details are complex. A 2005 United Nations Food and Agriculture Organization (FAO) report deplored the "myth" that deforestation causes floods. According to the report, conservationists often blame major flood events on deforestation. For example, floods on the Brahmaputra River in Bangladesh are often attributed to deforestation in Nepal, a major

FIGURE 5.2. The border between Haiti and the Dominican Republic shows a vastly greater amount of deforestation in Haiti (on the left). Satellite image courtesy of NASA/Goddard Space Flight Center Scientific Visualization Studio.

watershed that feeds the river. The FAO report denounces such broad generalizations.[2]

Their statement, however, was an unfortunate overstatement. Experts in hydrology have never claimed that deforestation always leads to flooding or that forests always prevent floods. Experts recognize several reasons for this. First, if a very large amount of rain falls, as during a heavy and prolonged summer monsoon season, the soil becomes saturated and will no longer hold water. The additional rain contributes to floods—trees or no trees. Second, major rivers are fed by tributaries, each with their own watershed. Each of these watersheds may experience a different pattern of rainfall, and have a different history of deforestation. The flooding in one watershed may be evened out by the lack of flooding in other watersheds of a major river system. The relationship between rainfall, forest cover, and flooding also depends on many geographical factors, such as the steepness of the mountain slopes. Third, in most parts of the world, floods are controlled by reservoirs and dams. (However, floods occur when the reservoir capacity is exceeded, rather than immediately after heavy rains.) Deforestation causes soil erosion, which fills up the reservoir with silt; a silted reservoir cannot hold back as much flood water. Fourth, it is difficult to prove a correlation between deforestation and flooding over large regions and long periods of time. While deforestation is occurring, the climate is undergoing natural variability, and the greenhouse effect is also causing climate patterns to change (chapter 3). Natural climatic variability, along with the greenhouse effect, might either cancel or intensify the flooding attributable to deforestation. Finally, the economic and human impact of flooding depends on how the people downstream are living. In Bangladesh, floods nearly always have tremendous impact, as many poor people have no place to live except on the floodplain.

Nevertheless, most studies confirm that deforestation increases runoff and flooding. Nearly all studies that have compared intact and deforested watersheds, on a relatively small scale (a few square miles or less), have demonstrated that flooding increases when forests are cut down.[3] Perhaps the most famous experimental confirmation of this idea took place at the Hubbard Brook Experimental Forest in New Hampshire. Two adjacent watersheds, on the slope of the same mountain, had very similar vegetation and geological structure. Researchers cut down all of the trees on one of the watersheds. Almost immediately, more water ran off of the clear-cut watershed than the intact watershed (fig. 5.3). As the forest grew back on the clear-cut slope, the difference between the cut and intact watersheds decreased.[4] Another long-term study began in 1941 on the Coweeta watershed in western North Carolina. Forests were clear cut on some slopes and left intact on others. In this study as well, deforestation increased stream flow.[5] Researchers have concluded that deforestation can also increase flooding over large regions.[6] Computer simulations also consistently conclude that deforestation increases flooding.[7]

Instead of making a general claim that deforestation does or does not cause flooding, it would be more accurate to say that deforestation increases the *risk of* flooding and of mudslides *in local regions*. No report has disputed this. Although deforestation in Nepal may not have been

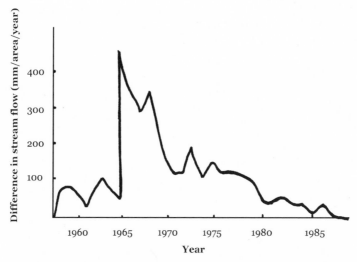

FIGURE 5.3. The difference in runoff between a clear-cut and an intact watershed at the Hubbard Brook Experimental Forest in New Hampshire. When one watershed was clear-cut in 1965, it immediately had more runoff than the undamaged forest; as forest on the clear-cut slope recovered, the difference in runoff between the two slopes decreased. See note 4. Illustration by the author.

the principal factor that caused flooding in Bangladesh, it has certainly contributed to flooding and mudslides in Nepal.

It has been popular among environmental writers to compare forests to sponges, because they absorb water when rain is abundant and may release it gradually when rain is absent. Most of the water that flows in rivers comes from groundwater. If forests enhance the penetration of rain into the groundwater, then they should also promote the flow of water in rivers during dry seasons. If this is the case, then a bare hillside presents the humans that live at its base with the worst of both worlds: floods during the rainy season, and drought during the dry season. Whereas there is a great deal of evidence that forests often reduce the risk of flooding, it is much less clear that they enhance the flow of water in rivers during dry seasons.[8]

A major reason that forests may not consistently augment the volume of river water is that they use much of the water for their own transpiration (chapter 4). Some trees that live in arid zones have very high rates of transpiration. By transpiring so much in a land without much rain, these trees can actually reduce the amount of rainfall that recharges the groundwater and the rivers. Eurasian tamarisks, for example, have filled many thousands of miles of floodplain along rivers in the southwestern United States; their prodigious capacity for transpiration severely depletes stream flow. Eastern red cedar is a native shrub that is invasive in many pastures of the south Midwestern plains of the United States, and its transpiration reduces the availability of water for livestock.

A watershed covered by managed forests, by perennial crops, or a combination of the two (as in agroforestry), usually does not control flooding as well as a native forest. But there are exceptions. In the North Carolina watershed study cited above, slopes that were planted with white pines after clear-cutting actually had *less* runoff than intact slopes, principally because the evergreen pines slowed down the rainfall all year, including in the winter when the deciduous trees had no leaves. But tree plantations do not control flooding and soil erosion unless they have a ground cover of small plants.[9] Many Japanese cedar plantations, established early in the twentieth century and without good ground cover, have far greater soil erosion and water runoff than do natural forests.[10] Tree plantations have been used in China for decades as a way of rehabilitating degraded land. But when these plantations consist of single species and do not have adequate ground cover, they are not effective at preventing erosion and flooding. In Fujian Province, native forests have been cut down and replaced with commercial plantations of pine and eucalyptus. In these plantations, runoff and flooding have increased, because the native forests had deeper soils that could retain more water.[11]

It is therefore unfortunate that the lack of precise correlation between deforestation and flooding has caused economic and government leaders in many countries to dismiss the importance of conserving forests and of replanting those that have been cut—and now they can cite the FAO report as justification for that position. However, it was not the intent of the FAO report to undermine forest conservation or reforestation. The main concern of the report is that there is a danger in blaming major floods on deforestation. In many countries with a lot of poor people, much deforestation is caused when poor people gather firewood and clear forests for subsistence farming. Often, cutting the forest is their only option for survival. However, the governments of these countries can simply issue a logging ban—which they may or may not seriously expect to be enforced—and thus create the appearance that they are solving the problem.

Forests and other native vegetation can also help to reduce waterlogging in soils. In many low-lying areas, the water table is salty. If the water table rises too much, this saline water can reach the roots of the plants and kill them. Forests allow rainwater and snowmelt to percolate into mountain soils, rather than rushing down to the plains. This reduces the chance that the saline water table in the lowlands will come too close to the surface.

Although our understanding of the water cycle and the importance of plants in it is a relatively recent scientific development, humans have been observing these processes for millennia, with varying degrees of comprehension. In *Critias*, Plato noted the connection between deforestation and the runoff of rain water: "In the primitive state of the country its mountains were high hills covered with soil . . . and there was abundance of wood in the mountains. Of this last the traces still remain, for although some of the mountains now only afford sustenance to bees, not so very long ago there were still to be seen roofs of timber cut from trees growing there . . . and there were many other high trees. . . . Moreover the land reaped the benefit of the annual rainfall, not as now losing the water which flows off the bare earth into the sea." He then added: "What now remains compared with what then existed is like the skeleton of a sick man, all the fat and soft earth having wasted away, and only the bare framework of the land being left."[12] Observers ever since Plato, including early conservationist George Perkins Marsh in the nineteenth century, have agreed.

Therefore, as rain falls and begins its headlong rush to the sea, forests (and to a lesser but still significant extent, shrublands and grasslands) slow the water down, preventing floods and mudslides. We have it exactly backward when we cut down the forests and then pray for rescue from the floods and droughts.

The Water Goes Back Up

There is yet another way in which forests help to prevent droughts besides diverting rain into the ground rather than letting it run off. Everyone knows that forests need rain. In places where rain is scarce, so are trees. However, the equation works, to a lesser but nevertheless important degree, in the opposite direction as well: forests influence rainfall because transpiration increases the humidity of the air. Much of this water condenses to form clouds, which can then precipitate rain back onto the forest. Trees, therefore, recycle some of the rainwater rather than allow all of it to flow downstream to the ocean. In this way, plants depend on the rain, and the rain depends on plants.

As noted in chapter 4, the amount of water transpired into the air by plants is enormous. The transpiration from some forests can alter the climate. The Great Smoky Mountains in North America are not smoky, but misty, partly because of the water vapor transpired by its thick forests. Scientists estimate that a quarter of the rain that falls on these Appalachian forests comes, via transpiration, from the forests themselves. Fully half of the rain that falls on the Amazonian rain forest is recycled from the transpiration of these forests.[13] On a worldwide scale, from one-quarter to one-third of the water that enters forests and grasslands comes out as evaporation, much of which is transpiration from plants.[14]

Now consider what happens when forests are cut down and not replanted. Transpiration stops, except for water vapor that transpires from the few weeds that may replace them. The water flows down to the sea, perhaps as floods. Once in the ocean, some of the water returns as rain, because ocean water evaporates to form clouds, which drift back over the land. The amount of rain, however, will be less than before, because the rain that once came from both transpiration and from oceanic rain clouds now comes only from the latter (fig. 5.1). Deforestation, therefore, sometimes creates a drier climate.

Destruction of the native plant cover, and the resulting soil erosion, may have caused climates to become drier throughout human history. Forests formerly covered much of the now-semiarid Mediterranean region. The Anatolian peninsula (now Turkey) used to have extensive forests, but they have been cut down. Massive soil erosion has filled in the bays, so that the site of ancient Troy, which was once near the ocean, is now several miles inland.[15] Today, there is simply not enough rain in the Mediterranean region to support the growth of large trees. The major cause was probably a natural climatic shift that would have occurred anyway. But deforestation enhanced the impact of the drier climate. A similar effect may have occurred on some of the islands of the eastern Atlantic. Madeira is Portuguese for "woodland," which is what covered it until Portuguese colonists cut many of the forests down to raise sugarcane. "Verde" is Spanish for "green," which is what the Cape Verde Islands were until Spaniards cut down many of their forests for sugarcane. The Spaniards also destroyed the forests of the Canary Islands, as well as driving the native Guanche people into extinction. Forests have grown back on these islands, but apparently they are drier and sparser than was the case before European colonization. This observation cannot be proved for certain, but Christopher Columbus noted concerning the Canary Islands, Madeira, and the Azores: "since the removal of forests that once covered those islands, they do not have so much mist and rain as before."[16] The relative contributions of natural climate change and deforestation cannot be disentangled, but deforestation must have been an important factor. The deforestation of North America, which has had a wetter climate than the Mediterranean for perhaps thousands of years, did not result in a drier climate. This suggests that the effect of deforestation on rainfall is greatest in locations, such as the Mediterranean, in which precipitation is already relatively low.

Perhaps the most famous example comes from the southeastern Pacific island that the Polynesians called Rapa Nui. Polynesian explorers found this island covered with forests, including the world's largest species of palm. They established a thriving society. Their ruling clans competed with one another to build gigantic stone heads, which were rolled from quarry to coastal cliffs on tree trunks. The trees were depleted—for agriculture, for fuel, and for this monumental building program. By the time Europeans encountered this island in the eighteenth century, on an Easter Sunday (thus its modern name

Easter Island), the forests were gone, agriculture had failed, and the inhabitants were starving.[17] Today, the island has no native forests and few trees of any kind. There is not enough rain for them.

It may seem incredible that deforestation could affect rainfall on a single island. Rainfall on islands, it would seem, must be overwhelmingly determined by the vast oceans around them. However, in these cases, deforestation may not have reduced rainfall very much but rather hindered the ability of the island's surface to retain the water, either in the soil or in the form of fog and mists. So, although deforestation might have been one reason that the climate became drier on Easter Island, it does not explain the whole story. The society disintegrated primarily because of soil erosion, not because of the drier climate.[18] Recent evidence indicates that the forests disappeared not just because humans cut them down, but also because rats (brought by humans) consumed the seeds of the palm trees, which were the dominant forest component.[19] The collapse of Easter Island has been cited as a microcosm of what may be happening to the whole world.[20]

Today, destruction of plant cover may be causing drier climate in Australia. In 1907, Australians built a 2,000-mile fence to keep rabbits, which had been unwisely introduced from Europe, out of the farmlands (which it failed to do). The fence today divides agricultural land from native shrublands. Thirty-two million acres of native vegetation have been made into cropland. On the agricultural side of the barrier, rainfall has declined by 20 percent, whereas cloud formation appears to be more common on the wild side of the fence.[21] Other examples are plentiful. Computer simulations indicate that deforestation will reduce rainfall in the Amazon rain forest.[22] And deforestation is already causing the cloud layer to form at a higher altitude in Central America. These clouds formerly blanketed extensive "cloud forests" whose plants and animals depend on the continuous mist, but now they are often left high and dry, and some have died.[23]

Deforestation does not always lead to drier conditions, of course, nor can all droughts be attributed to it. The Sahel region of Africa has not experienced enough deforestation to account for the long dry spells from which it continues to suffer.[24] In other areas, such as India, the reduction in rainfall that has resulted from deforestation has been compensated by the increased evaporation that irrigation promotes. That is, less water vapor comes from trees, and more from irrigated fields. At the present time, on a worldwide scale, evaporation of water has decreased by 4 percent as a result of deforestation, a reduction of 1,850 cubic miles of liquid water equivalent per year. Irrigation has increased evaporation by 1,600 cubic miles per year. Therefore loss of natural plant cover may not cause reduced rainfall in India, where there is a lot of irrigation, but it might cause droughts in the Amazon and in parts of Africa.[25]

The reason that there is no simple correlation between deforestation and rainfall is that rainfall is influenced by many factors. For example, shifting ocean currents cause the El Niño Southern Oscillation (in which warm El Niño periods alternate with cool La Niña periods). Even the orbit of the earth and volcanic dust can influence climate. Therefore, scientific research cannot prove a cause-and-effect relationship between deforestation and reduced rainfall—not in the above-cited examples from the past nor in the world today. The magnitude of transpiration is great enough, however, that deforestation must have an effect on rainfall.

Because deforestation can cause a reduction in rainfall, a positive feedback loop can result. The reduction in rainfall causes the decline of the remaining forests, owing to drought or to fire, and inhibits the regeneration of forests. Such a positive feedback can cause rain forests such as the Amazon to vanish more quickly than would be suggested by current rates of deforestation. Deforestation reduces rainfall, and this further reduces the forest.

From the East Indies to the Mediterranean to Polynesia, humans have destroyed forests and in so doing have altered the climate. Soil erosion resulting from agriculture has also caused the collapse of civilizations. In the Babylonian *Epic of Gilgamesh*, King Gilgamesh offends the god Enlil, who causes the kingdom's farms and rivers to fail. But it may have been damage to the forests that caused the environmental catastrophes immortalized in this epic tale. In the biblical story, Adam and Eve saw four rivers watering the Garden. They could not have suspected that the Garden itself may have made the flow of water in the rivers possible.

Water and Salt

Plants remove water from the soil and transpire it into the air. The cells of the sapwood of trees are the incredible conduits of millions of tons of water each year. Not all of the water that enters a plant transpires into the air. Each living cell of the plant contains a pool of water. Like the cells of animals, plant cells absorb water by accumulating dissolved molecules such as salt and sugar and amino acids. But unlike the cells of animals, plant cells accumulate enough dissolved substances, and therefore draw in enough water, to create water pressure. Each plant cell has a strong cell wall that prevents it from exploding. It is this pressure that allows soft leaves to hold themselves upright; if this pressure is lost, the leaves wilt. In some cases, for example, small plants on a humid night, this pressure can actually push water up through the stem (fig. 5.4).

Plants can readily absorb water from the soil only if their cells have more dissolved materials than the soil. When salt accumulates in the soil, the ability of the plant to take up water is reduced. It is difficult for roots to remove water from salty soil for the same reason that drinking salty water makes you thirsty: water will diffuse out of the blood into the stomach, or from the plant into the soil, rather than the other way around. This situation is sometimes called "physiological drought": the water is present, but not available ("Water, water, everywhere, but not a drop to drink," as Samuel Taylor Coleridge's Ancient Mariner said). Plants that evolved to live in salty environments, such as salt marshes, have special adaptations that allow them to deal with physiological drought.

Crop plants (with the exception of barley and sugar beets) cannot tolerate even slightly salty soil. But salty soil is what crops often encounter in irrigated fields. This occurs when the water with which the field is irrigated evaporates, leaving behind an accumulation of salt, a process called *salinization*. Salt can build up to high levels over the course of decades and centuries. The water of the lower Colorado River, which is heavily exploited for irrigation, contains a ton of salt per acre foot of water. Irrigation water with less salt can also cause salinization, even though it may take longer for a significant buildup to occur. Salinization can also occur if salty water accumulates in the soil.

FIGURE 5.4. Pressure buildup in roots can push water out of leaves. This occurs only in small plants on warm, humid nights. Photograph by the author.

Salinization is extensive in some countries (table 5.2).[26] Evaporation occurs most rapidly in hot, dry regions, which is exactly where irrigation is most intensively practiced, such as California and Australia.[27] An overall world average is 10 to 15 percent. Perhaps the most striking and visual example of salinization has occurred in the Aral Sea between Kazakhstan and Uzbekistan of the former Soviet Union. Soviet-era centralized agricultural plans demanded that much of the flow of the Amu Darya and Syr Darya rivers be diverted to irrigate cotton plantations. This had three effects. First, much less water entered the Aral Sea; as a result, it began to dry up. Its surface area was reduced by half

TABLE 5.2. Percentage of Irrigated Farmland Affected by
Salinization in Selected Countries

COUNTRY	PERCENTAGE OF FARMLAND AFFECTED
Egypt	37
Iran	29
Kenya	29
Kuwait	86
Libya	40
Nigeria	34
Turkmenistan	46
United States	23

SOURCES: United Nations Food and Agriculture Organization at
http://www.fao.org/nr/water/aquastat/regions/Africa/index.stm and
http://www.fao.org/docrep/W4356E/w4356e06.HTM.

in just a five-year period between 1960 and 1965. The collapse of the fishing industry in the Aral Sea is graphically portrayed by ships rusting in the middle of what is now barren desert but was once one of the world's largest inland seas. Second, severe salinization afflicted the farmlands. Not only did this reduce the very yields that irrigation was intended to enhance, but salty dust created lung disease in the local populations. Finally, because the little bit of water that made it into the Aral Sea (which was originally more of a lake than a sea) was salty, the salinity of the water that remained there tripled. The public health problems in the Aral Sea region were not very important to the central government in Moscow; this was the same government that dumped highly radioactive waste in open ponds in Uzbekistan. Now that these countries are independent, however, restoring the Aral Sea and rescuing the agricultural lands from salinization has become a priority.[28]

Salinization of agricultural soil has been going on for thousands of years. A 1958 study claimed that ancient Mesopotamian records show a gradual decline in agricultural yields and that the yield of barley relative to wheat increased over the course of two millennia.[29] The authors considered this evidence of salinization, because barley is more salt-tolerant than wheat. The earliest city-states were near the Tigris-Euphrates delta; over time, the center of political power moved upstream, as downstream agricultural soils probably experienced salt buildup. Although this interpretation is likely to be correct, the shift from wheat to barley in the ancient Near East may have been in response to drought rather than salt. Because of its more rapid life cycle, barley tolerates drought better than wheat.[30]

Salinization has a considerable economic impact. The Food and Agriculture Organization (FAO) of the United Nations estimates that salinization can cost $625 per acre in reduced yields, which amounts to $11 billion worldwide.[31] Certain techniques have been developed that can reduce salinization. One of these is to use a great deal of irrigation water and let it quickly drain. This technique works only if there is a lot of water available for irrigation, which is frequently not the case. Moreover, this approach can cause waterlogging, which may make salinization worse if saltwater rises in the soil to the level of the roots. Among the more promising techniques to reduce salinization is drip irrigation. This method applies water directly to the soil over the root-ing zones of the plants, rather than to the entire soil surface. This not only helps to conserve scarce water resources but also reduces surface evaporation. Another technique is the use of ground cover, often made of plastic. This cover not only helps to reduce surface evaporation (of water often supplied by drip irrigation) but also restricts the growth of weeds, which would not only compete with the crop plants but transpire much of the water from the soil. The crop plants grow through a narrow slit in the plastic. This procedure is very effective but also very expensive, so it is most frequently used on high-value crops like strawberries. Worldwide adoption of techniques such as drip irrigation will be necessary if irrigated agriculture is to continue.[32]

Forests are not just a pretty green carpet on the landscape. They slow down the processes by which the landscape washes away, which can be disastrous for the cities perched precariously on it. Forests even out the flow of water, and they moderate the climate. They create a livable world not only in terms of oxygen (chapter 2), global temperature (chapter 3), and local temperature (chapter 4), but also in terms of water. The forests and fields of this lovely planet help to protect us from floods and droughts.

chapter six

PLANTS FEED THE WORLD

*I have often thought that if heaven had given
me my choice of position and calling, it should
have been on a rich spot of earth, well watered,
and near a good market for the productions of
the garden. No occupation is so delightful to me
as the culture of the earth, and no culture
comparable to that of the garden. . . . I am
still devoted to the garden. But though an
old man, I am but a young gardener.*
—THOMAS JEFFERSON

The rise of fast food chains—spreading late in the twentieth century from the United States to every corner of the world—is one of the defining characteristics of contemporary society. Anywhere in the world you go, you can get exactly the same kind of hamburger and fries, untouched by the finger of any local proprietor's creativity. And it is, or seems, cheap. As Eric Schlosser has shown in his book *Fast Food Nation*, the boom in fast foods exacts a tremendous cost on an unskilled and underpaid workforce.[1] Morgan Spurlock has explained how a diet of fast food can lead to numerous health problems.[2] Given the societal and health-care costs, fast food may not be such a bargain. But our modern food production system has given us food that, at least, seems inexpensive.

Although a hamburger may not cost much money, the *energy* cost of the hamburger is enormous. The energy in the food itself comes from the sun. Most of this biological energy is lost as it moves through the industrial food chain, from corn to cow to human. Even more than this,

FIGURE 6.1. This roadside sign proclaims the efficiency of modern agriculture. (The number has since risen to 129.) But modern agriculture is efficient only in terms of labor; it is very wasteful of energy. Photograph by the author.

our food production system relies on the prodigious expenditure of fossil fuels. Modern technology allows American agriculture to be very efficient in terms of the amount of food produced by a small number of farmers (fig. 6.1). Cheap modern food and productive modern farmers require the use of enormous amounts of fossil fuel energy.

How Plants Make Food

Of all the sunlight energy that shines on the earth, 99 percent just heats the earth's surface—the rocks, the water, and the plants. Sunlight energy drives the engine of the weather and climate. It not only warms the air but also evaporates water to form clouds and bring rain. When leaves absorb sunlight and become hot, they use transpiration to cool themselves off and thereby create cool shade from which we can benefit (chapter 4). The remaining 1 percent goes into photosynthesis and produces all the food in the world. This is the same process by which plants put oxygen into the air and remove carbon dioxide from it (chapters 2, 3).

Photosynthesis uses sunlight energy to synthesize food molecules. More specifically, it brings together energy from the sun and small molecules from the air and soil to create food molecules. Plants use energy from sunlight and electrons from water to bind together carbon dioxide molecules, producing carbohydrates such as sugar. Plants make all of their other molecules from these carbohydrates. Plants and animals need food to provide energy to run their bodily activities and molecules from which to build their tissues. The food molecules created by plants uniquely fulfill both of these needs. As science writer Michael Pollan says, plants create life out of thin air; another science writer, Natalie Angier, says that plants spin the sun into gold and make sweetness from light.[3]

First, I will outline what happens to food energy as it goes through the food chain. Then I will consider the fossil fuel energy that is consumed by the human food chain.

Food Energy: The Tyranny of the Food Chain

Some people eat mostly vegetables and grains, while others eat a lot of meat. The calories of energy in food come from photosynthesis either way. So what difference does it make whether you get your food energy from meat, or from vegetables and grains? The answer is, 90 percent. Here's how I arrive at that number. Consider a herd of cows eating grass. They munch on the above-ground greens, like natural lawnmowers, but they do not eat the roots. They do not digest all that they eat; some of it comes out as waste, which feeds the decomposers, such as bacteria and fungi, in the soil. They use some of the molecules that they digest for their own metabolism (even though cows are not particularly energetic, they do need energy for walking, mooing, and producing body heat). Only a small amount (typically less than 10 percent) of the energy that was originally present in the grass will end up being incorporated into the tissues of the cows. Now consider humans eating cows. We do not eat the entire cow: we eat the muscles as meat, but neither the bones nor the hide (opinions differ regarding the edibility of other parts, such as tripe). Further, we do not digest all that we eat. And we use some of the energy that was in the meat as the source of our own metabolism, for movement and for body heat production. Only about 10 percent of the energy originally present in the cow is incorporated into human tissues.

This simple truth about the food chain is known as the Ten Percent Law. Only about 10 percent of the energy and molecules from one level of the food chain is incorporated into the next level. It is as true for wild food chains (such as birds eating insects that eat plants) as it is for human ones. In fact, 10 percent is more or less an upper limit: frequently far less than that makes it from one link in the food chain to the next. One of the early demonstrations of the Ten Percent Law was a 1942 study by ecologist Ray Lindeman. He determined the energy content of all of the producers (photosynthetic organisms), herbivores (plant eaters), and carnivores in Cedar Lake Bog in Minnesota.[4] Lindeman chose to study a bog because it was relatively self-contained, exchanging little material with surrounding areas. The amount of energy in each level of the food chain (*trophic level*) never exceeded ten10 percent of the energy in the level on which it depended (fig. 6.2).

A tremendous amount of American farmland is used to produce feed for cattle and other livestock. Seventy-five million acres of prime American farmland are used to raise grains that are fed to livestock.[5] As you drive through the vast agricultural landscapes of Iowa and Illinois, most of what you see is corn and soybeans for livestock. The amount of American farmland that grows grain to feed cows and produce enough beef to feed *one person* could feed *ten people* if the farmland was used to produce grains and vegetables that people can eat directly, eliminating the intermediary cow in the food chain. A hundred pounds of corn produces 10 pounds of beef, which produces 1 pound of human. At most, *1 percent* (10 percent of 10 percent) of the food energy in the corn reaches the person, via the cow. If, instead, we used our agricultural land to raise grains and vegetables for humans, 100 pounds of corn would

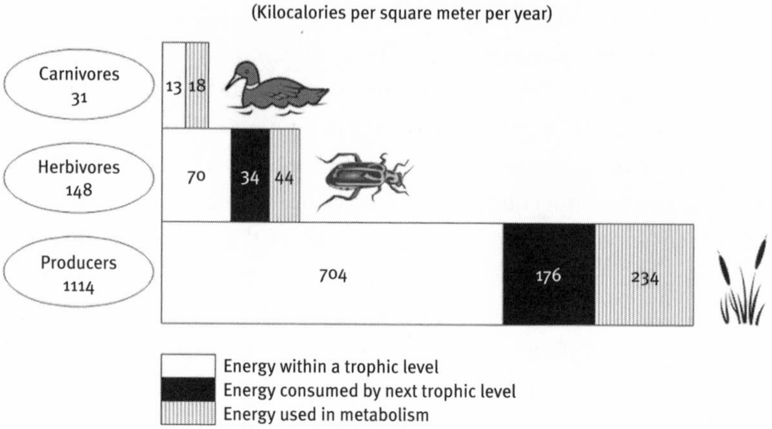

(Kilocalories per square meter per year)

FIGURE 6.2. In one of the earliest studies of the flow of energy through the food chain (in Minnesota's Cedar Lake bog), Ray Lindeman found that usually less than 10 percent of the energy in one level of the food chain is incorporated into the next level. See note 4. Illustration by the author.

produce 10 pounds of human. This is why the difference between calories from plants and calories from livestock is 90 percent.

Cornell University ecologist David Pimentel estimates that American farmland could feed 800 million vegetarians.[6] Around the world, billions of people are malnourished, leaving them susceptible to diseases that they would otherwise have resisted. Although few people actually die of starvation except in times of war, many die of malnutrition that leads to disease. One reason the world has so many hungry people is that rich consumers demand, and farmers produce, grain-fed animal products from confinement operations. There is not much money to be made in raising grains for poor people to eat.

Livestock producers like to point out that the efficiency of beef, pork, chicken, egg, and milk production is considerably higher than 10 percent. A hundred pounds of cattle feed, in a confined animal feeding operation, can produce more than 10 pounds of cow. It takes 3 to 4 pounds of feed to produce a pound of beef just prior to slaughter, a feed conversion efficiency of about 30 percent. Chickens are even better: it takes only about 2 pounds of feed to produce 1 pound of chicken. It would therefore seem, at first, that modern agricultural technology has freed us from the tyranny of the

[114]

Ten Percent Law. This is not, however, the case. The calculations apply only to feed *from the bag*, not from the ground. The feed in the bag represents agricultural plants that have already been harvested, leaving behind the chaff, and processed to remove indigestible parts. Although 100 pounds of cattle feed can produce about 30 pounds of cow, 100 pounds of corn plant (roots and all) will produce far less than 10 pounds of beef. Moreover, in some calculations, the animal product is usually fresh weight of meat or egg, full of water, while the feed is relatively dry. The difference is made up when the animal visits the watering trough. Ignoring water weight in the calculation makes the feed conversion efficiency artificially high.

Animal breeders are always trying to improve feed conversion efficiency. Some breeds of livestock produce less waste than, or convert their food energy to marketable products better than, other breeds. One extreme example is genetically engineered sheep in Australia, whose intestinal bacteria produce less methane: this leaves more food for the production of wool. The gains in feed conversion efficiency have been economically important to livestock producers, but are very small in comparison with the Ten Percent Law.

Livestock production does not need to be so expensive in terms of photosynthetic energy. First, although we use 75 million acres of farmland to raise feed grain, we also use 680 million acres of pasture. Pasture, in general, is land in which the soil is too poor or too dry to produce crops. Grass-fed beef, therefore, comes to the consumer without the calculation of guilt (as long as the pasture is not overgrazed). The Ten Percent Law still applies, but it applies to grass that humans cannot eat. Second, it is not necessary to use grain and oilseed that are raised on farms to produce livestock feed. Traditionally, swine have been raised on slop, or leftover kitchen waste. This is still done in backyard pigsties around the world. But industrial-scale hog production uses feeds that come from the farm, largely because kitchen waste would need to be sterilized to prevent the spread of bacterial disease.

It is not necessary to become a vegetarian in order to "eat lower on the food chain," making more efficient use of the calories produced by farmland. The average American eats more than 60 pounds of beef,

50 pounds of pork, and 100 pounds of poultry a year. Nobody is purely carnivorous, and relatively few people are purely vegetarian. If we shifted our agriculture away from livestock feed and toward food for humans to eat, our efficiency would rise from a little over 1 percent to a little under 10 percent. We could achieve this and still enjoy eating meat—by using meat as a luxury rather than a staple in our diets.

The Price of a Hamburger

The above calculations barely begin to reveal the true price of meat production. Ancient sunlight energy trapped mainly inside of petroleum is used in prodigious quantities at every stage of bringing meat to the table. I will again use the corn-to-cattle-to-human example.

First, consider the corn production stage. Corn seeds, which were processed and preserved by seed companies, are delivered to farmers in bags by truck. Seed processing, packaging, and transportation require energy from petroleum. Farmers use large equipment and a great deal of diesel to plow the fields and to plant the seeds. They also use lots of chemical fertilizer. Fertilizer production, such as processing trona mineral ore into potassium fertilizer, requires much energy. Ammonia (nitrogen) fertilizer is produced directly from atmospheric nitrogen gas and natural gas by an industrial synthesis known as the Haber-Bosch process. Most of the nitrogen that crops absorb from the soil comes from this process; therefore most of the nitrogen atoms in your body come from this process. It takes a lot of energy to package and transport the fertilizers to the farmers. Farmers use large equipment, and even more diesel, to apply fertilizer to their fields. Modern farmers use databases and global positioning devices to apply just the amount of fertilizer that they need in each location, which saves on fertilizer and minimizes water contamination; but it still takes energy. They also use pesticides and herbicides and equipment to control insects and weeds in their fields. Producing and delivering these chemicals takes even more energy. The availability of herbicide-resistant crops has reduced the use of mechanical weed control but has increased the use of herbicides, which contribute to environmental toxicity. In many areas, maximum corn yield requires irrigation. Pumping water from the ground, especially in regions where irrigation has drawn the water table down to 100 or more feet below the surface, requires energy. And finally, farmers use large equipment and lots of diesel to harvest the corn and separate the seeds from the stalks. It takes even more energy to transport the corn seeds to locations where they are processed and bagged as cattle feed.

Now consider the beef production stage. Energy is required to deliver cattle feed to the confinement operation. It takes a lot of diesel to transport young calves from the fields where they begin their lives to the finishing operations where they are fed grotesquely large amounts of corn, which is not their natural food. The maintenance of the confinement operation requires a lot of energy to deliver the food and water, to remove the wastes, and to administer antibiotics, which may be the only way to prevent the spread of disease in the crowded, filthy conditions of the confinement operation. It takes even more diesel to transport the finished cows to the slaughterhouse, to operate the slaughterhouse, and clean up the wastes. Processing, refrigeration, and transportation of the meat hundreds of miles to the fast food facility or to the supermarket requires even more energy. Nearly every food item in a supermarket or a restaurant (not just the beef) is transported by trucks a great distance from the place that it is manufactured—about 1,500 miles on the average.

It is not surprising, then, that livestock products from the industrial food chain require a huge expenditure of fossil fuels, calorie for calorie. Once again, chicken production is relatively cheap, requiring only 4 calories of fossil fuel to produce 1 calorie of chicken meat. But beef, according to David Pimentel, requires 54 calories of fossil fuel per calorie of meat that humans eat (table 6.1). And this is just the meat. A similarly huge expenditure of energy occurs in the production of the buns from wheat, and the condiments that make up the American hamburger meal.

All of this energy expenditure results in a tremendous amount of carbon dioxide production, which contributes to the greenhouse effect (chapter 3). In addition, cattle and other livestock release methane, an

TABLE 6.1. Fossil Fuel Calories required per Calorie of
Animal Product.

ANIMAL PRODUCT	CALORIES OF FOSSIL FUEL
Cattle	54
Lamb	50
Eggs	26
Pork	17
Milk	14
Turkey	13
Chicken	4

SOURCE: David Pimentel, "Livestock Production: Energy Inputs
and the Environment," *Canadian Society of Animal Science
Proceedings* 47 (1997): 17–26.

even more potent greenhouse gas, from their digestive systems. It is safe
to say that the average consumer has no idea how much energy, and
how much carbon emission, results from the products that he or she
purchases. If products were labeled with this information, many consumers would choose their purchases at least partly on the basis of it,
and would buy fewer products. Although this would reduce sales, it
would help the world. Tesco, the largest supermarket in the United
Kingdom, plans to label all of its seventy thousand products with the
information about the amount of carbon generated by their production,
transport, and consumption.

Considering all of the fossil fuels used in this process, how can a
hamburger possibly be cheap? Part of the answer is that the U.S.
government supports the industrial food chain in two important ways.
First, it directly subsidizes corn production. Government payments
enable farmers to grow corn despite the tremendous costs. As a result,
we produced more corn than we knew what to do with. Science writer
Michael Pollan said that corn was a solution in search of a problem.[7]
What do we do with the corn? We feed it to cows. Beef is cheap because
corn is cheap. Inexpensive corn allows the production of high-fructose
corn syrup and corn starch. These processed ingredients appear in a
startlingly large number of processed food products. The hamburger
meal contains corn-fed beef, but also buns and mayonnaise and mustard
and ketchup that contain ingredients made from corn. During blizzards
in Colorado in December 2006, military aircraft brought hay to

stranded cows, which cost about $1,000 per cow. Modern agricultural products are cheap in the economic balance because the U.S. federal government sits on the scales. Second, the government provides not only money for farmers but military force to guarantee the availability of cheap fuel. We import a lot of petroleum. We exercise our military interest in the Middle East to ensure the continued flow of petroleum into world markets. Agriculture uses a larger amount of petroleum products than do automobiles.

Corn, however, is no longer a cheap commodity in search of a market. Increasingly, we have distilled corn into ethanol as an alternative fuel for transportation, in an attempt to reduce the amount of gasoline that we use. It is true that ethanol contains more chemical energy than is required to produce it. But when we consider the amount of fossil fuel that is required to raise the corn, we are using more fossil fuel ·by producing ethanol than we are saving. Ethanol producers can pay more for corn than livestock producers had been paying, and much more than poor people who need it for food can pay. By 2008, a great deal of the corn produced in the United States was being purchased for ethanol production. The federally mandated and subsidized production of ethanol caused the price of corn to increase. At the same time, soaring oil prices caused the production of corn, and many other crops, to become even more expensive. The result was a global food crisis in 2008, because poor people could not afford the high prices. U.S. government subsidies have also made it difficult for farmers in other countries to sell their produce. A farmer in Mexico can produce corn more cheaply than can an American farmer, but federal subsidies make American corn cheaper than Mexican corn. Many Mexican farmers have now migrated north to become migrant farm workers in the United States.[8]

There are many ways in which the expenditure of fossil fuels can be reduced in the process of farming. As mentioned above, annual plowing and the application of fertilizers are major energy expenses in agriculture. Agricultural systems that avoid annual plowing and fertilizers can eliminate these energy costs. No such agricultural systems exist on a large scale today, but they are being developed by diligent research. This research receives very little support from the federal government and is certainly not supported by industry; it receives its primary funding from private donations. The Land Institute, in Salina, Kansas, conducts research into energy-efficient and environmentally safe agriculture.[9] Staff scientists and graduate student interns from programs around the United States are developing what they call natural systems agriculture (NSA).[10] Two of the major characteristics of NSA are the use of perennial crops and the mixture of different kinds of crops. Nearly all modern crops, such as wheat and corn, are *annuals*: they live for only one year and must be replanted every year. To raise these crops, the farmer must plow the ground each year. *Perennial* crops, on the other hand, maintain a living rootstock from one year to the next and do not have to be replanted. The Land Institute is developing perennial crops that may someday replace the annual crops on which agriculture currently depends (see chapter 11).[11] The use of a mixture of crops can also save energy relative to the planting of a single crop. Crops require nutrients such as nitrogen. In a field consisting solely of corn, the nitrogen must be supplied by the farmer. But if strips of corn alternate with strips of alfalfa, the alfalfa can provide nitrogen to the corn (see below). The use of crop mixtures in which one of the crops supplies nitrogen to the other can make the expensive application of nitrogen fertilizer unnecessary.

With all of the subsidies, it would appear that farmers should be very wealthy. Everybody knows that this is not the case. Remember that the corn that the farmers produce must be sold as inexpensively as possible. Larger farming operations, which use more petroleum energy, can produce corn at a lower cost than small farms. Farmers have little choice: they must either sell their corn cheaply, or not sell it at all. And they must produce as much as possible, even though this lowers the price of corn, because only the biggest producers survive in the market.

One way to reduce the expenditure of fossil fuels in modern agriculture is to encourage smaller farms. Government subsidies strongly favor the largest farms. But small farms are often more productive, acre per acre, than large ones. One reason for this is that farmers more carefully cultivate each acre of their land when their farms are small. The reason that large farms dominate the agricultural market is not because they are more productive relative to their size but because of the distribution of their produce. Transportation of agricultural produce from farm to processing, and from processing to market, requires a lot of energy and money. It is the business of specialized corporations, not of the farmers themselves. A corporation that buys from farmers would much rather buy a big haul of produce from one 1,000-acre farm than from ten 100-acre farms: it would get the same amount of produce with one-tenth the effort. It is the economies of scale, made possible by the expenditure of fossil fuels, that makes large farms more profitable than the small farms that are often more productive and sustainable.

Consumers can encourage the success of small farms by purchasing produce at farmers' markets, thereby eliminating the distributors that favor large farms.[12] This is only the beginning of the advantages that follow from the direct patronage of small-farm produce. Crops from small farms are often organically grown. The consumers often get food that may be of higher nutritional quality and that has fewer pesticide residues.[13] The elimination of the middleman saves up to 90 percent of the energy involved in transportation, with its attendant carbon emissions. Buying local produce usually, but not always, reduces a consumer's carbon footprint.[14] There are, of course, some disadvantages to buying local produce. You cannot simply buy anything you want anytime you want it. Part of the 90 percent energy savings is the fuel that would have been used to truck winter produce up from Mexico into the United States. Our expectations will need to be less extravagant. As Michael Pollan has written, small-scale agriculture requires not just a new kind of farmer but a new kind of eater as well. But the benefits would include food security: we can have food that does not need to be transported from long distances regardless of petroleum costs or availability. As Barbara Kingsolver says, "we could make for ourselves a safer nation, overnight, simply by giving more support to our local food economies and learning ways of eating and living around a table that reflects the calendar."[15]

As if the direct advantages were not enough, farmers' markets (and, to a lesser extent, natural food stores) have immense social benefits. Pollan also explains that our food production system depends on consumers knowing nothing about the food except its price. But going to a farmer's market can be an educational experience. Nowhere else do consumers get to talk to and learn directly from food producers. Surveys show that consumers are ten times more likely to talk to one another in a farmer's market than in a supermarket. The promotion of community is an unplanned side-effect of participation in small-scale food production. It will take a lot of people going to a lot of farmers' markets, however, to bring about any substantial change in the current hegemony of large farms.

Food Molecules

Photosynthesis produces sugar, but plants make many other kinds of molecules out of that sugar. They convert sugar directly into the more complex carbohydrates. They store food energy in starch molecules and use cellulose fibers to strengthen their tissues. They also use sugar to produce lipids in their seeds, such as the oils that they use for energy storage, and the waxes that coat the outer surfaces of leaves. Plants also absorb minerals such as phosphorus from the soil, which they use along with sugar to make many of their complex molecules. Plants consist of thousands of kinds of molecules, all of which begin as sugar.

Many of these molecules, which are full of stored sunlight energy, can be used not only by the plants but by humans, who release the energy by burning wood. Coal, oil, and natural gas have resulted from the compression and chemical transformation of dead plants and animals that piled up in swamps up to 300 million years ago. Therefore fossil fuel energy also comes from plants.

Animals depend on plants for all of their molecules. As just one example, animals need proteins. Plants can make proteins from sugars, but animals cannot. Therefore animals obtain protein, as well as sugar, from the plants that they eat (or from the animals that ate the plants). Therefore, by absorbing carbon dioxide from the air and minerals from the soil, and by being food for animals, plants do something that no other large organisms can do: they put carbon, oxygen, and minerals into the food chain (fig. 6.3).

The nitrogen cycle is remarkable because nitrogen atoms move through both the atmosphere and the soil. The soil component of the system begins when plant roots remove ammonia, nitrates, and nitrites from the soil and incorporate them into molecules such as proteins. After nitrogen atoms pass through the food chain, decomposers release them back into the soil. The atmospheric component begins when certain bacteria (the *nitrogen-fixing bacteria*) take nitrogen gas from the atmosphere and convert it into ammonia; these bacteria therefore operate as fertilizer factories. Some are photosynthetic, forming pond scum. Other nitrogen-fixing bacteria are not photosynthetic and live in the soil. They obtain nitrogen gas from the air that penetrates into the soil from the atmosphere. But green plants play an important role even in the activities of nitrogen-fixing bacteria. Many of these bacteria live in special swellings or *nodules* in the roots of plants. The plants feed them with sugar and proteins. They even protect the bacteria from excessive exposure to oxygen by producing a variety of hemoglobin. Such plants therefore maintain miniature fertilizer factories in their roots. More nitrogen-fixing bacteria live in root nodules than out in the soil. Legumes, such as beans, alfalfa, clover, and acacia, maintain colonies of bacteria such as *Rhizobium* in their nodules, whereas alder trees maintain colonies of *Frankia* bacteria. When leguminous plants shed leaves or die, the nitrogen enters the soil, where it can enter the food chain again. Other bacteria, the *denitrifying bacteria*, transform nitrogen compounds back into nitrogen gas, which diffuses from the soil into the atmosphere (fig. 6.3).

This is one of the reasons why natural systems agriculture (see above) can promote soil fertility without the addition of fertilizers. If leguminous plants, such as alfalfa, grow together with other crop plants such as grains, the alfalfa can fertilize the soil and promote the growth of the grain crops.[16]

Plants Produce Many Other Materials

Humans depend on the molecules produced by plants not only for food but for almost everything else. Plants produce all of these materials for

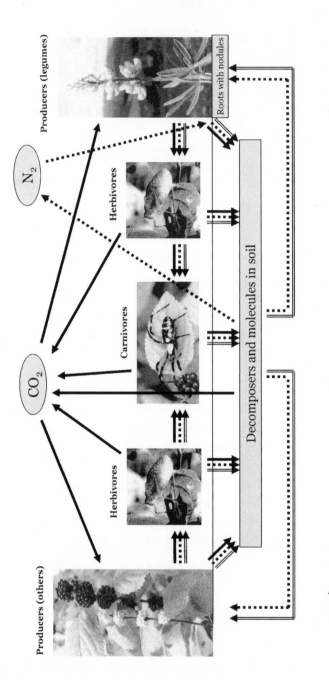

FIGURE 6.3. Plants play an essential role in the cycles of all kinds of atoms. The arrows represent the movement of carbon, nitrogen, and minerals (such as phosphorus or potassium) in the food chain. Illustration by the author.

their own use, but they happen to be useful to the human economy as well. Here are just a few examples.

Fiber and wood. Plants produce textiles and fibers—wood that we make into paper, as well as cotton, flax, hemp, jute, and sisal that we use for fabrics and ropes. Trees produce wood that is still the superior material for many construction purposes and for furniture.

Industrial materials. Despite the availability of synthetic plastics and rubber, cellophane is made primarily from wood cellulose, and natural rubber (from the sap of rubber trees) remains in great demand. Synthetic colors have not completely replaced natural plant dyes such as indigo, henna, and tannins. The fruits of the jojoba bush, native to the deserts of southwestern North America, produce a liquid wax that is very stable under high temperature and pressure, making it valuable for industrial applications. Some insecticides, such as pyrethrum, are still made from plants. Even synthetic industrial materials come largely from plants. They are made from petroleum, which is compressed and chemically converted plants that died in swamps millions of years ago.

Pharmaceutical agents. Although synthetic medicines may now dominate the market, many pharmaceutical chemicals (such as morphine, codeine, digitalin, and atropine) come from plants, and most others, even though now produced synthetically (such as quinine, which often cures malaria, and aspirin, the most widely used medicine in the world) were first discovered in plants. The anticancer drugs vincristine and vinblastine were first discovered in the leaves of a little pink wildflower in Madagascar. Plants synthesize these molecules in order to defend themselves from herbivores, and these molecules happen to also have medicinal effects when humans use them in small quantities.

Food components. Natural food additives such as gums and laxatives usually come from plants. Some chemicals from plants, while bitter and dangerous in large quantities, act as food supplements when eaten in small quantities, such as caffeine in coffee, theobromine in chocolate, tannic acids and essential oils in tea, and the vast panorama of spices.

Plants not only feed the world but also supply the human economy with essential materials. Because of plants, eating is an interesting and exciting activity, rather than just the ingestion of sustenance. Is there anything that we do in life that is not made possible by the growth of plants? In this and previous chapters, therefore, we have discovered that plants allow us to breathe, drink, and eat.

chapter seven

PLANTS CREATE SOIL

The nation that destroys its soil destroys itself.
—FRANKLIN DELANO ROOSEVELT

During the Dust Bowl of the late 1920s and 1930s, millions of tons of topsoil from the western great plains of the United States blew for thousands of miles in the air, leaving a withered and almost sterile substrate that would grow very little food or forage. In many midwestern towns, midday became as black as night, and some people even died from the inhalation of dust. This was at the end of a long drought, but the dust storms were not completely the fault of the drought. They were also the result of overgrazing and excessive plowing, which left the soil that could otherwise have survived a drought vulnerable to the relentless wind.

In April 1935, during the height of the Dust Bowl, soil scientist Hugh Bennett met with the U.S. Senate Public Lands Committee. He wanted to convince them to create a government agency to combat soil erosion. When the clouds of dust blew over Washington D.C., he invited the senators to look out the window. They saw the nation's farmlands, upon which both the present and the future of the country depended, blowing out to sea. The Senate established the Soil Conservation Service (now the Natural Resources Conservation Service) of the U.S. Department of Agriculture on April 27, 1935. This agency helps farmers to incorporate new techniques that will reduce the loss of topsoil to the forces of wind and rain.

We all know that plants need soil. The purpose of this chapter is to convince you that soil needs plants. Soil is not dirt. Soil contains dirt, but it is a world teeming with organisms, on which the aboveground world utterly depends. Plants grow out of the soil and feed on it. However, it was also plants that created the soil, and that continue to feed it. There was a time when the earth had no soil. Before plants began to live on land about 400 million years ago, the continents had plenty of dirt, but no soil. There was sand, there was silt, there was clay; but there was no humus, therefore there was nothing for soil organisms to eat. It was dead. Then terrestrial plants evolved, grew their roots into the dirt, and transformed it into the living material called soil. Plants can create soil in a devastated landscape. We desperately need for them to render us this service.

A Tour of the Soil

Soil is permeated by billions of tiny passageways that allow oxygen from the atmosphere to reach the plant roots and the soil organisms that need it. Carbon dioxide from the respiration of roots and soil organisms diffuses out of these galleries into the atmosphere. If you were extremely tiny, and standing on the surface of the soil, you could walk right into the soil through these tunnels. If you did so, you would see both inorganic and organic components (fig. 7.1).

Inorganic Components

After entering the soil, you would see tiny fragments of rock. The largest fragments are called sand, and the fragments smaller than sand are called silt. The smallest particles are clay. Your journey would be uncomfortable, because these clay particles would be sharp, like billions of tiny blades. Clay particles are not simply smaller than silt and sand; they have been chemically transformed by the weathering effects of water and oxygen. As a result of this, they have a negative electrical charge on their surfaces. The negative charges attract positively charged atoms and molecules. Many of these atoms and molecules are inorganic nutrients that plants require for their metabolism and growth. Potassium allows plants

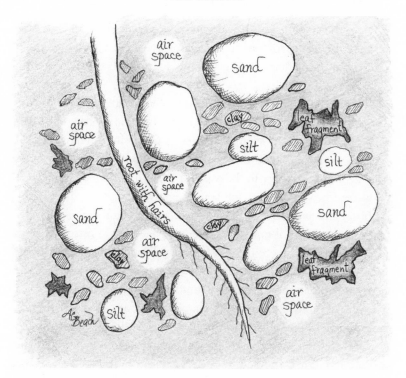

FIGURE 7.1. Components of healthy soil include inorganic particles and organic materials. Illustration by Gleny Beach.

to maintain the water balance of their cells, just as your body requires sodium. Calcium glues plant cells together. Magnesium is a component of chlorophyll. Ammonium is a form of nitrogen from which plants manufacture their proteins. These atoms and molecules are dissolved in the water that is in the soil.

If you continued your walk into the soil, however, you would notice that the water clings to the soil particles, especially to the clay. There must be a force that is holding the water onto the soil particles, counteracting the force of gravity. This force is, in fact, the negative charges on the clay surfaces. Water molecules are electrically charged; although the positive and negative charges balance one another out, each water molecule has two positive poles and a negative pole. Water therefore clings to the clay surfaces because of its positive poles. When it rains, some of the water adheres to the clay surfaces, while the remaining water drains through the passageways. Plant roots absorb the water that clings to the soil particles.

Therefore, clay allows soil to hold on to both water and to positively charged minerals. Plant roots absorb both of them from the soil. When roots absorb positively charged minerals from the water in the soil, minerals from the clay surface replace them. So plants not only obtain minerals from the water in the soil, but the soil renews the minerals in the water. The soil also contains some negatively charged minerals, such as phosphates and nitrates. In fact, these are among the most abundant and important of soil mineral nutrients. They remain dissolved in the water but do not cling to the clay surfaces. Because clay surfaces do not recharge nitrogen and phosphorus, soils tend to become depleted of them sooner than they do of potassium and calcium.

Hydrogen ions are small but pugnacious. They displace the positively charged minerals from the soil surfaces, allowing them to wash away. Hydrogen ions are the source of acid; therefore, acidic soils such as those found in bogs tend to be very poor in nutrients because their clay particles cannot hold them.

Organic Components

On this microscopic journey you would also see many organic components of the soil, some living, some formerly living. The dead components include fragments of leaf litter and other plant parts. Owing to their size and irregular shape, they do not fit tightly together. It is largely because of leaf litter fragments that the soil remains spongy and contains passageways that allow air and water to penetrate. Soil that is compacted not only lacks sufficient oxygen but may be so hard that plant roots cannot penetrate it. *Humus* is leaf litter that has decomposed so much that its structure is no longer recognizable but still contains many complex, and sticky, organic molecules. Humus makes the soil brown. It also glues the soil particles together, reducing their tendency to erode. The organic matter of the soil also helps to hold on to soil nutrients. Therefore, a soil rich in organic matter is a soil in which plants can grow well. Modern agriculture pays attention to the mineral components of the soil, and billions of dollars are spent on inorganic fertilizers. Organic fertilizers, from mulch to manure, not only supply many of the same minerals, but also enhance the ability of the soil to retain these minerals.

Other organic components include living organisms. Bacterial cells and strands of fungi are abundant, busily transforming leaf litter into humus and humus into mineral nutrients such as nitrates, phosphates, and potassium. Many other microorganisms and small invertebrate animals live in the soil. Perhaps most important are the earthworms. A soil without earthworms is likely to be nothing more than dirt, for earthworms are both an indicator and a maintainer of soil health. Earthworms eat soil and digest organic materials from it; therefore a lack of earthworms indicates a lack of organic matter. They maintain soil health by creating burrows, through which air and plant roots can penetrate.

Plants are the world's champion miners. They absorb billions of tons of minerals from the soil and draw them up through their roots and stems into their leaves, where they use them to produce their molecules. Plant roots absorb more iron from the soil than is extracted by humans from all the iron ore in the world. But whereas humans permanently deplete the supply of minerals, plants merely borrow them. When the plant finishes using them, it sheds its leaves and allows the decomposers to return the minerals to the soil.

The Tragedy of Erosion

We have already seen how the destruction of vegetation leads to both drought and floods, and may alter the local climate. Destruction of vegetation can, in addition, allow wind and rain to erode much of the soil away and destroy the fertility of what remains. Soil erosion not only destroys fertility but also the ability of the soil to hold water. Eroded soil is also more vulnerable to the impact of drought.[1] All over the world, soil erosion

[131]

is occurring at alarming rates, far beyond the ability of natural processes to replace the soil. The Food and Agriculture Organization (FAO) of the United Nations estimates that one-half of 1 percent of farmland is lost each year to erosion, with the result that 9 percent of the world's land surface has been degraded by agriculture (table 7.1).[2] Worldwide, available cropland per person has decreased from 1.3 acres in 1960 to just 0.6 acres in 2006. Agricultural productivity per hectare has increased during that time, with the result that this decline is seldom noticed. Eventually, the intensity of agricultural production will reach a maximum, and if soil erosion continues, food production will decline.

TABLE 7.1. Percentage of Land Area Degraded by Agricultural Activities in Selected Regions and Countries.

REGIONS/COUNTRIES	PERCENTAGE OF LAND DEGRADED
Africa	8
Burundi	65
Nigeria	75
Somalia	21
Asia and Pacific	12
China	22
Indonesia	11
South Korea	42
Europe	11
Germany	21
Romania	46
United Kingdom	19
Near East	6
Iraq	33
Syria	22
North America	12
Canada	2
United States	23
North Asia	6
South and Central America	9
Brazil	11
Cuba	25
Mexico	22
Nicaragua	65
world	9

SOURCE: United Nations Food and Agriculture Organization, "Terrastat," available at http://www.fao.org/ag/agl/agll/terrastat.

[132]

In the United States, soil erodes faster than it is formed on 40 percent of the farmland. In many places the loss of soil—owing both to the outright loss of agricultural acreage and to reduced fertility of the acreage we continue to use—is canceling out the progress made by improved crop breeding and the use of artificial fertilizers. Almost every farmer today understands the problem; every rural county is organized into a soil conservation district; and no one today would boast of having "worn out" three farms, as did some late-nineteenth-century cotton farmers. The degree of cooperation between farmers and soil scientists is high, considering that compliance of farmers with soil conservation guidelines is totally voluntary. Despite this, soil erosion continues in the United States.

Soil erosion has been a problem since soon after the beginning of agriculture. Early, small-scale farms caused little erosion. In fact, all over the world, farmers developed soil conservation measures such as terracing. In wet areas of the Americas, such as the Amazon rain forest and the swamps of northern Florida, farmers even built up soil by mixing compost with shells, bones, and broken pottery. When botanist William Bartram traveled through northern Florida in the late eighteenth century, he found many sites, where Native American villages had once stood, in which the soil had been artificially built up (chapter 1).

As the scale of agriculture increased and soil conservation practices were discontinued, soil erosion became a major problem in the ancient world. Deposits of silt thousands of years old indicate that extensive soil erosion occurred in ancient Israel, Greece, Italy, and Spain. Soil erosion has contributed greatly to the collapse of past civilizations.[3] Two major centers of prehistoric Native American agriculture—the Anasazi of the Mesa Verde vicinity, and the Hohokam of what is today Phoenix—abandoned large acreages that even had advanced irrigation systems. Their civilizations collapsed hundreds of years before the arrival of Europeans (chapter 1).

Soil erosion not only causes a problem where the soil is lost but also where it accumulates. Minerals from the eroded soil can end up in lakes and promote the explosive growth of algae. Soil erosion fills harbors, lakes, and reservoirs with silt. In some areas of China, with severe erosion, dam-building projects had to be abandoned soon after construction because of siltation of the reservoirs. Indeed, every reservoir has a finite life span, calculated at no more than a couple of hundred years.

Plants to the Rescue

Agriculture is one of the major causes of soil erosion. This occurs because annual crops die at the end of the growing season and must be replanted at the beginning of the next season. In between harvest and planting, and during the time when the seeds are germinating and the seedlings are small, the soil may be virtually barren and vulnerable to erosion. But this does not need to be the case. As mentioned in the previous chapter, researchers at the Land Institute in Salina, Kansas, are developing natural systems agriculture (NSA), which is based on perennial crops.[4] Perennial crops do not die at the end of the growing season and do not need to be replanted. There is no time of the year in which the soil is unprotected. Perhaps the major way in which perennial crops protect the soil is through their astonishingly deep root systems. Wheat has very shallow roots and protects the soil from erosion very ineffectively, even during the growing season. Wild wheatgrass, a perennial relative of wheat, has much deeper roots, and the roots hold the soil in place all year long (fig. 7.2). If agriculture was based on perennial rather than annual crops, not only would agriculture use less energy (see previous chapter), but it would lose less soil.

Plants do not just need soil, but fertile soil, in which to grow. When the soil does not contain enough mineral nutrients, the farmer must add them. As explained in the previous chapter, supplying fertilizer to the soil is one of the major expenses of energy and money in agriculture. In the past, much of the fertilizer came from farm animals such as horses and cows. The farm animals ate some of the crops, such as oats and alfalfa, and their wastes were used as fertilizer. In modern agriculture, farms seldom have animals, so the farmer must buy fertilizer; and in confined animal feeding operations, there is nothing for which the wastes can be used. A harmonious system, in which plants feed animals and animals fertilize plants on the same farm, has been replaced by a system in which fertilizer

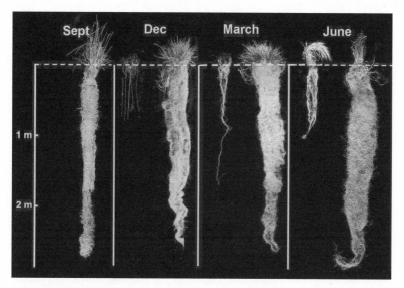

FIGURE 7.2. Wild wheatgrass has a much deeper root system than does annual wheat. Photograph courtesy of Jerry Glover, Land Institute.

must be purchased and wastes become a source of pollution. As writer Wendell Berry put it, modern agriculture has divided a solution into two problems.[5]

Despite this tremendous expense, much of the fertilizer that farmers add to the soil is wasted. It is difficult for farm equipment to supply fertilizer to the soil in the middle of the growing season, when the crop plants are large; therefore, farmers usually apply fertilizer before planting. Much of this fertilizer washes away, or (in the case of nitrogen) is lost into the air, before the growing plants have a chance to use it. This is as true for organic fertilizers, such as cattle or pig manure, as it is for inorganic fertilizers such as ammonia. If you can smell the manure on a field in the springtime, then at least some of the fertilizer is going into the air, not into the soil.

Researchers at the Land Institute are also developing systems in which two kinds of crops can be grown together, and one of them supplies fertilizer to the other. Staff researcher Dave Van Tassel showed me an arrangement in which strips of alfalfa grow in a field of sunflowers.

Alfalfa is a leguminous plant in which bacteria manufacture nitrogen fertilizer (see previous chapter). The old leaves and stems of the alfalfa decompose and release nitrogen on the top of the soil, while old roots die and release nitrogen below the surface; from both sources, the sunflowers receive a dose of nitrogen fertilizer. The advantage of this system is that nitrogen release occurs in the middle of the growing season, which is precisely when the sunflowers need it but at a time when mechanical fertilizer application would be difficult. Such a system could be used for many other crops besides sunflowers. The decomposing plant material promotes soil fertility not only by reducing erosion and by adding nutrients but also by enriching the soil with humus. The humus, in turn, allows the soil to hold more water than soil that gets its fertility primarily from a fertilizer bag. The Land Institute is also breeding perennial relatives of the major annual crop plants (see chapter 11). Large-scale experiments by Land Institute researcher Jerry Glover show that perennial crops, although not as good as wild prairie plants, can promote soil fertility and water-holding capacity much better than annual crops.

There is yet another way in which plant roots may help us rescue our soils from destruction. In many places, industrial pollution has contaminated the soil with toxic materials that make it unsafe for people to live on or near them, or drink the water that drains from them. What can we do with these soils? Many contaminated sites have been designated by the U.S. government as "Superfund sites," which will be cleaned up by a partnership between private industry and the federal government at some unspecified time in the future. Our choices seem rather bleak: we either surround these sites with fences and stay away from them forever, or else we try to remove the contaminants. But it would take millions of dollars to clean the soils, and when we were finished, we would have a slurry of mud—mud from which the toxins have been removed, but in which nothing will grow, because it will harden into brick. Is there some way to clean the soil without destroying its structure?

Several species of wild plants have specialized in absorbing toxic minerals from the soil. They use these minerals to protect themselves from herbivorous animals. They store the toxic minerals (such as selenium) in compartments known as vacuoles in their leaf cells, so that the toxins will not come in contact with their own cellular machinery, and their own cells are not harmed by them. However, when an animal eats the leaf, the vacuoles burst open and poison the food. Cattle that eat "locoweed" (genus *Astragalus*) become crazy or "loco" because of the effects of selenium on their nerves.

These plants are not trying to clean up the soil by removing selenium from it, but this is precisely the effect that they have. If we grow plants on a toxic waste dump, and these plants are specialized to remove that particular kind of toxic mineral, they can clean the soil for us. Their tiny roots will penetrate and clean the soil without disturbing it. Researchers at major universities are using genetic engineering to produce breeds of plants that remove toxic minerals from the soil at a higher rate than the natural plants can.[6] Once the plants have grown, we can harvest their stems and leaves, full of the toxic material, and dispose of them in a safe place.

These plants will not only clean up the dirt that had been barren and eroded because of its toxicity; they will also begin to add organic material to it, and transform it into soil. Just as the first land plants, long ago, transformed the dirt of a barren landscape into soil, they can do so again.

chapter eight

PLANTS CREATE HABITATS

When the poor and the needy seek water,
And there is none,
And their tongues are parched with thirst,
I the Lord will answer them,
I the God of Israel will not forsake them.
I will open rivers on the bare heights,
And fountains in the midst of valleys,
I will make the wilderness a pool of water,
And the dry lands springs of water.
I will put in the wilderness
The cedar, the myrtle, the acacia, and the olive;
I will set in the desert the cypress, the plane,
and the pine together,
That men may see and know,
May consider and understand together,
That the hand of the Lord has done this,
That the Holy One of Israel has created it.
—ISAIAH 41:17–20

In 1978, I hiked to the top of the tallest peak in the continental United States, Mt. Whitney in California, more than 14,000 feet above sea level. I looked toward the southeast down into Death Valley, which at more than 300 feet below sea level is the lowest part of the continental United States. Three years later, I visited Badwater, at the bottom of Death Valley, also in California. At both of these extremes, there were few plants: a few tufts of wildflowers (such as sky pilot *Polemonium*) huddled near the bases of rocks near the mountaintop, and a few pickle-weeds (*Salicornia*) grew in the extremely salty fluid at the bottom of Badwater. The alpine plants could withstand the freezing winds of winter, the desert plants could withstand salt and extreme heat, and both of them could withstand drought that would kill most other species.

Plants grow almost everywhere. There are no large regions of the earth, with the exception of the middle of Antarctica or the islands near the North Pole, that are totally barren. Even around the edge of Antarctica there are two species of flowering plants: the hairgrass *Deschampsia antarctica*, and the pearlwort *Colobanthus quitensis*, a distant relative of the carnation. You will probably never travel to a place that is truly lifeless and "God-forsaken," because wherever you go, plants will already be there.

Wild habitats are places where many thousands of species of organisms, from bacteria to bears, are found. But habitats are more than this. They are not merely places where organisms live but places that organisms *create*. And it is mainly plants that create these habitats. A forest is not merely a place that receives enough rainfall for trees to grow; it *is* the trees and other organisms. If it were not for the differences in the types of dominant plant cover, all parts of the earth would look more or less the same: rocks, sand, silt, and clay at different temperatures and with varying amounts of moisture. It is mainly the dominant plants that make Denmark, Afghanistan, Burma, and Brazil recognizably different places. As we have observed in previous chapters, plants make food, create and hold down the soil, put water vapor into the air, and produce shade in which other species (including other plants) live. They began doing so more than 400 million years ago, when the first land plants evolved.[1] In this chapter we will see how plants create many different kinds of habitats.

The broad pattern of plant geography on Earth results mainly from plant responses to temperature and moisture. They do so in three ways. First, there is a greater amount of photosynthesis and therefore plant growth in regions that are warmer and wetter (fig. 8.1). Second, there are more plant species in regions that are warmer and wetter. Third, the geographical pattern of forests, shrublands, and deserts results mainly from temperature and rainfall (fig. 8.2).

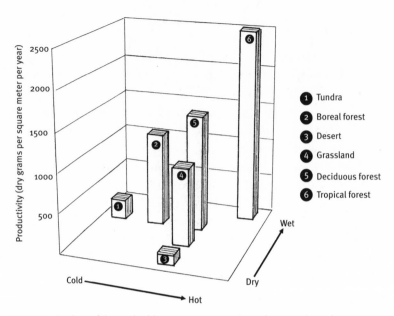

FIGURE 8.1. Regions of the earth with warmer temperatures and more moisture have a greater annual photosynthetic productivity. Illustration by the author.

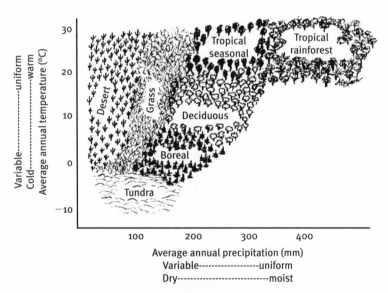

FIGURE 8.2. Temperature and moisture conditions determine the types of plant habitats over broad regions of the earth. Most grasslands are delimited by fire, as well as by temperature and moisture; grasslands are discussed in chapter 9. Illustration by Gleny Beach and the author.

Vegetation zones do not have distinct borders. You cannot draw a line on a map and say that this is where one vegetation zone ends and another begins. Perhaps the best example is the transition zone between tundra and forest, described below. The timberline of a mountain, which separates the forest from the tundra, is actually a zone. The same is true of all of the other borderlands between the vegetation zones described in this chapter. Vegetation zones are also not internally uniform. Each contains a diversity of microclimates (see below) and contains patchworks of disturbances (chapter 9).

Habitats and the Plants That Create Them

Our first journey takes us along a temperature gradient, from the coldest places to the warmest regions that have plenty of moisture: from tundra and boreal/subalpine forests, to temperate deciduous and coniferous woodlands, to tropical forests. The second journey takes us along a moisture gradient from the wettest to the driest regions within the zone of moderate temperature in North America: from western coniferous forests to deserts. Although these journeys are focused on North America, the principles that we learn from them will be the same in other parts of the world.

Tundra

One a summer afternoon I hiked higher and higher into the Sangre de Christo Mountains of Colorado. As I ascended, I saw smaller and smaller trees, until finally there were no trees at all. I had reached the alpine tundra, where the vegetation forms a flat green blanket. Tundra plants live in the coldest regions of the world, except those regions too cold to support any plants at all. The *arctic* tundra is in northern latitudes (such as Alaska and Canada), and the *alpine* tundra is on the tops of the tallest mountains (such as the Rocky Mountains and Sierra Nevada) throughout the continent. The most easily accessible alpine

[141]

TABLE 8.1. Locations and Dominant Plants of the Arctic and Alpine Tundra.

	EXAMPLES OF IMPORTANT SPECIES	WHERE TO FIND THEM[a]
Arctic tundra	Arctic birch (*Betula nana*) Arctic rose (*Dryas octopetala*) Arctic willow (*Salix reticulata*) Sedge (*Eriophorum vaginatum*)	Arctic NWR
Alpine tundra	Arctic rose (*Dryas octopetala*) Bistort (*Polygonum viviparum*) Sky pilot (*Polemonium viscosum*)	Rocky Mountain NP Sequoia NP Yellowstone NP

NOTE: NP = National Park; NWR = National Wildlife Refuge.

[a]Examples of locations easily accessible to travelers.

tundras are in Rocky Mountain National Park and in and around Yellowstone National Park (table 8.1). The alpine tundras are remnants of the arctic tundra, isolated on mountaintops after the glaciers of the most recent ice age retreated.

Having the world's lowest average temperatures means that tundras have both long, cold winters and brief, chilly summers. At first I felt warm as I stood in the Colorado tundra, in the early summer sun; a few moments later, however, a chilly rainstorm with a few flecks of snow descended on me. Snow can come any time of year in the tundra. Because the average annual temperatures are so low, tundra plants grow very slowly. Most of them (such as the grasslike sedges) are perennials, which store up food throughout each year of their slow growth. During the brief summer, tundra plants quickly come to life and put the warm weather to productive use. The tundra around me was ablaze with the colors of phlox, bistort, and gentians. The flowers all bloom in late summer because there is no other time warm enough.

The main thing you would notice about the tundra is that it has no trees. At least, not any trees that stand upright. If you look closely, you may see bushes and small trees, such as willows, alders, and birches, the stems of which actually trail along the ground. This occurs because the winters are so severe that nothing can survive in the tundra unless it is protected by a layer of snow. Occasionally a tree trunk emerges higher than the snow and is badly damaged as a result. The only branches that survive on these "flag trees" are on the leeward side of the trunk. Other trees hunker down in areas protected by rocks (fig. 8.3). The small, bushlike trees are called *krummholz*, a German word for "twisted wood." Winter survival is one reason that tundra plants are all small.

FIGURE 8.3. This whitebark pine tree on Lassen Peak at the timberline in California can survive the winter only in an area protected from the wind. Photograph by the author.

Another reason that tundra plants are small is that the soil is very shallow. The summer is so short that the soil thaws out only a few inches from the surface. Below this depth, the water in the soil remains permanently frozen, forming a layer of *permafrost*. Permafrost might as well be rock, because plant roots cannot penetrate it. Plants can grow tall only if their roots can grow deep. The shallow soil is, moreover, very poor. Rich soil contains nutrients released mostly by the decomposition of dead leaves. Tundra summers are short and chilly. The slow decomposition that results produces poor soil. Despite the low productivity of the tundra, its low vegetation can feed large herds of mammals such as musk oxen and caribou.

In some drier spots in the tundra, wildflowers occur in the form of cushion plants, such as the dwarf azalea. Composed of several different species, cushion plants really look like little cushions or pillows on the ground. This condensed growth form protects the stems from the effects of the wind and allows dead leaves to build up a rich soil directly underneath the plants. They therefore create tiny zones of good growing conditions immediately around them.

Even though the total amount of snow and rain that the tundra receives during a year may be relatively small, the tundra has an abundance of moisture. Arctic tundra is virtually flooded, and to walk through one you must step from one tussock of vegetation to another. Water accumulates in the flatlands of the arctic tundra because the permafrost does not allow the melted water to drain. Hordes of mosquitoes emerge in summer. One group of scientists quantified the mosquitoes by the "swat test"—the number of mosquitoes they could kill with one slap of the hand. The record (achieved no doubt by a scientist with big hands) was 270.[2] In the alpine tundra, most of the water drains from the high mountain slopes. However, substantial ponds of water can still accumulate, and mosquitoes are a problem for hikers there also.

Despite their similarities in temperature conditions, arctic and alpine tundras experience very different sunlight conditions. In the arctic tundra, the winter has long and often continuous nights, whereas the summer has long days when the sun may not set for several weeks. In the alpine tundra, days and nights are of more equal length: in

Ecuador (where it is called *páramo*) and Papua New Guinea these regions experience approximately twelve-hour days all year.

The tundra plants themselves create the conditions that allow them to persist for centuries, and for the regeneration of new tundra plants. They also create microhabitats in which animals find shelter. In the fierce weather, the poor soils could wash and blow away, but plants cover and protect the soil.

Boreal and Subalpine Forests

In slightly warmer areas, forests grow instead of tundra. The coldest forests in the world—the ones farthest north and highest on the mountains—are woodlands of coniferous trees, often spruces. (There is little land area in the southern hemisphere suitable for the growth of such forests.) Although these forests also contain some deciduous species such as aspens and birches, most of the trees are evergreens, whose needlelike leaves persist for several or many years before falling off the tree. Across northern Europe, Siberia, and Canada stretch immense *boreal* (northern) forests just south of the arctic tundra, whereas immediately below the alpine tundra you find a mantle of *subalpine* forests. Boreal and subalpine forests are coniferous, but the converse is not always true: there are many pine forests that are neither boreal nor subalpine (see below). The subalpine forest communities are remnants of the boreal forest, isolated on mountains after the glaciers of the previous ice age retreated. You can find subalpine forests in the mountains of many western states, and in a few eastern mountains as well, in the United States (table 8.2). The last remnants of a subalpine spruce forest continue to thrive in the Black Hills of South Dakota. A small patch of fir forest on the top of Mount Mitchell in the Smoky Mountains National Park of Tennessee and North Carolina has nearly died from an insect infestation that was caused by acid rain and increasing temperatures.

The winters in boreal and subalpine regions are long and cold, and the summers are short and chilly, though less so than those of the tundra.

TABLE 8.2. Locations and Dominant Plants of Boreal and
        Subalpine Forests.

	EXAMPLES OF IMPORTANT SPECIES	WHERE TO FIND THEM[a]
Boreal forest	Spruce (*Picea glauca*)	Glacier NP
Subalpine spruce forest	Spruce (*Picea glauca*)	Rocky Mountain NP Yellowstone NP
Subalpine pine forest	Bristlecone (*Pinus longaeva*) Limber pine (*Pinus garryana*)	Ancient Bristlecone Preserve (Calif.) Sequoia NP

NOTE: NP = National Park.

[a]Examples of locations easily accessible to travelers.

The trees of the boreal and subalpine forests can survive the winter winds, which are less severe than those of the tundra, and therefore they grow above the layer of snow. The long, cold winters explain why boreal and subalpine forests consist mostly of narrowly conical, needle-leaved evergreens. First, the shape of the trees allows the heavy winter snows to easily slide off. Second, the waxy, needlelike leaves are not very efficient at photosynthesis, but they are drought-resistant. Drought resistance is not something a boreal or subalpine tree needs during the moist summer; it is an adaptation to the long winter, whose winds can freeze-dry unprotected tissue. Third, the needles are evergreen because the summer is too short for the trees to grow a whole new set of leaves each year. They would use up more food to produce an annual new set of leaves than these leaves would produce. Rather, the leaves must live for several years. They are adapted more to surviving winter than to efficiently making food in summer. The needle form is inferior for photosynthesis but superior for overwintering.

Another reason that the leaves of boreal and subalpine trees persist for more than a year is that the soil is poor. Although there is no layer of permafrost near the surface of these soils, their cold temperatures allow little decomposition and therefore little fertility. Moreover, the needles themselves (the principal component of the organic litter) are acidic. The acid leaches nutrients from the soil. In boreal forests, leaching is so severe that an ash-gray layer of poor soil is produced right in the root zone. No wonder the trees grow slowly even in the summer—and that there is very little undergrowth in a boreal or subalpine forest. When a tree sheds its leaves, it loses whatever nutrients are contained within them. Evergreen trees hold on to their leaves, thereby retaining a supply of nutrients not readily available from the soil.

Not all subalpine forests are alike. Spruces dominate most North American subalpine woods. But on the drier slopes of the Rocky Mountains in North America, and in the drier mountains of California, scattered pines take precedence. Among these are the bristlecone pines. In the White Mountains of eastern California, bristlecone pines are famous for being the oldest organisms on the earth, with an age exceeding four thousand years (fig. 8.4). These widely spaced pines grow very slowly in poor, rocky soil, never quite dying. They are like giant bonsai trees.

The plant species diversity of boreal and subalpine trees is relatively low. Only one or a few species of trees may dominate an area, overshadowing a sparse understory of scattered wildflowers. Despite the low diversity of plant species, many mammal species, including moose, lynx, and arctic hares, live in the boreal forests. Coniferous trees grow and ameliorate the harsh climate, creating a forest habitat.

Deciduous Forests

As you continue into yet warmer regions, you encounter a different kind of woodland. Rich forests of broad-leaved deciduous trees such as maples, beeches, oaks, and hickories cover most of the eastern United States and southeastern Canada. The deciduous forests have shorter, warmer winters and longer, warmer summers than the boreal and subalpine forests. The longer summer explains why these forests are deciduous and why they are broadleaved.

The trees are deciduous because they can afford to make a new set of leaves each year. The summer is long enough that photosynthesis allows a leaf to pay for itself and more in one season; it is therefore expendable in the autumn. The senescence of leaves, in which many of the molecules are disassembled and stored in the stems, takes time. A long growing

FIGURE 8.4. This bristlecone pine in California (*Pinus longaeva*) is at least three thousand, and possibly four thousand, years old. During the brief, chilly, and dry summer, the trees produce a few new needles and a thin layer of new wood. The needles live for decades. Over the millennia, they have produced very thick trunks. New branches replace old branches, but the tree does not become taller. Photograph by the author.

season is thus necessary not only for photosynthesis but for the completion of senescence. The trees lose their leaves in the autumn because if they did not, the snow that would cling to them would be heavy enough to break the branches, something that does not happen to trees with needlelike leaves. Sometimes a freak early snowstorm occurs before the leaves fall. This is a time of disastrous tree—and power line—damage.

The leaves are broad and thin because this shape allows very efficient photosynthesis during the summer. The maximum amount of green tissue is spread out in broad leaves to collect the sun. So broadleaved plants can grow faster than needle-leaved plants, on the average. Broad leaves are an adaptation to rapid photosynthesis in the summer, whereas needle leaves are an adaptation to surviving the winter.

Another difference that any visitor to a coniferous and a deciduous forest will notice is that the deciduous forest has much more undergrowth: an abundant, and sometimes dense, growth of herbaceous wildflowers, shrubs, vines, and tree seedlings occurs on the forest floor. This is largely because of the richer soil of the deciduous forests, which in turn results from the warmer temperatures and greater decomposition of leaves and release of nutrients.

Because the growing season is longer, it is not necessary for all of the flowers of the deciduous forest floor to bloom at once, as they do in the tundra. On the deciduous forest floor, early spring wildflowers grow quickly in the bright sunlight as soon as the snow melts and before the trees leaf out. These spring *ephemerals* (so called because of their brief aboveground existence each year) such as spring beauties and trout lilies finish blooming and die back to the ground about the time that the leaves of the trees expand. Other wildflowers such as phlox grow more slowly and dominate the forest floor in the shady summer, while yet others such as asters bloom in the autumn.

Deciduous forests have a greater number of dominant tree species than do the boreal forests. There are several different kinds of deciduous forests in North America (table 8.3). Forests of beech and sugar maple grow in the low-lying and moist regions across the northern Midwest, including much of Ohio and Indiana. Forests with numerous species of oaks and hickories grow in the slightly drier regions, such as Missouri, and in the more mountainous regions, such as the Alleghany and Great Smoky Mountains of North America. Many of these forests were once dominated by American chestnut (*Castanea americana*), but a fungus has destroyed all of the large adult trees in eastern North America; many

TABLE 8.3. Locations and Dominant Plants of Deciduous Forests.

	EXAMPLES OF IMPORTANT SPECIES	WHERE TO FIND THEM[a]
Beech-maple forests	Beech (*Fagus grandifolia*)	Great Smoky Mountains NP
	Sugar maple (*Acer saccharum*)	Shades SP (Ind.)
	Tulip tree (*Liriodendron tulipifera*)	
Oak-hickory forests	American elm (*Ulmus americana*)	George Washington
	Bur oak (*Quercus macrocarpa*)	Carver NM
	Mockernut hickory (*Carya tomentosa*)	
	White oak (*Quercus alba*)	Buffalo NR
Cross Timbers	Blackjack oak (*Quercus marilandica*)	Turkey Mountain Urban Wilderness (Tulsa, Okla.)
	Post oak (*Quercus stellata*)	
	Winged elm (*Ulmus alata*)	

NOTE: NM = National Monument; NP = National Park; NR = National River; SP = State Park.

[a]Examples of locations easily accessible to travelers.

survive only by resprouting. The driest and hottest deciduous forests in North America are the Cross Timbers forests, found mainly in Oklahoma and dominated by blackjack and post oaks (fig. 8.5).

Small evergreen shrubs may grow in the understory of deciduous trees. Some hollies may keep their broad leaves all winter, because they are protected by the canopy above them from excessive snow. Like ephemerals, these understory evergreens carry out most of their photosynthesis in the early spring before the canopy leaves open. They also carry out photosynthesis in the autumn, while the canopy leaves are falling.

A greater variety of deciduous trees dominate the forests of the south than those of the north. As you travel further south, the winters get shorter and milder, until the point is reached at which some broadleaf tree species do not lose their leaves at all. Woodlands in the southern portions of the United States contain some broad-leaved evergreens, such as magnolias (*Magnolia grandiflora*) and live-oaks (*Quercus virginiana*). The forests that are richest in species diversity in the United States are those of the Great Smoky Mountains. Winters and summers are mild. The amount of rainfall is very high. The trees absorb much of this rainfall through their roots and transpire it into the air through their leaves, producing mists that look like smoke. Practically all of the forest trees of the eastern United States can grow there, and most do.

FIGURE 8.5. This post oak tree (*Quercus stellata*), a dominant member of the Cross Timbers forest in Oklahoma, has grown slowly and produced twisted branches because of the dry, poor soil. Photograph by the author.

Deciduous forests also have many mammal species, from squirrels to bears. Some of these animals, such as deer, are as much at home in the yards of people who live near the forests as they are in the forests themselves. The trees create shady, moist conditions, and the rich soils that make the deciduous forest habitat what it is.

[151]

Moist Coniferous Forests

Some conifers grow in places that have climate conditions similar to those of deciduous forests. Some conifers (particularly pines) may specialize on relatively poor, rocky soils within the deciduous forest zone (see below). As we will see in the next chapter, other conifers grow quickly after disturbances, as do white pines (*P. strobus*) in the northeastern United States, the loblolly pine (*P. taeda*) of the southeastern United States, and the eastern red cedar (*Juniperus virginiana*).

But in other cases, coniferous forests live where they do because they survived the climate changes that have occurred in the past 30 million years—especially those of the ice ages (table 8.4). Thirty million years ago, there were many conifers in the relatively mild climate of Asia and North America. In North America, many of these have become extinct. The forests of giant sequoias in the Sierra Nevada of California and the coastal coniferous forests of the Pacific Northwest are also remnants of coniferous forests that were much more extensive in the past.

Giant sequoias, the largest trees in the world, are so immense that a typical suburban house could sit on one of their stumps. The largest

TABLE 8.4. Locations and Dominant Plants of Coniferous Forests.

	EXAMPLES OF IMPORTANT SPECIES	WHERE TO FIND THEM[a]
Moist coniferous forests		
Sierras:	Giant sequoia (*Sequoiadendron giganteum*)	Sequoia NP
	Incense cedar (*Calocedrus decurrens*)	Kings Canyon NP
	Sugar pine (*Pinus lambertiana*)	
Northwest:	Coast redwood (*Sequoia sempervirens*)	Redwood NP
	Douglas fir (*Pseudotsuga menziesii*)	Olympic NP
	Port Orford cedar (*Chamaecyparis lawsoniana*)	Redwood NP
Ponderosa forests	Douglas fir (*Pseudotsuga menziesii*)	Black Hills NF
	Ponderosa pine (*Pinus ponderosa*)	
Piñon-juniper woods	Juniper (*Juniperus osteosperma*)	Mesa Verde NP
	Piñon (*Pinus edulis*)	Grand Canyon NP

NOTE: NF = National Forest; NP = National Park.

[a]Examples of locations easily accessible to travelers.

of these trees, the General Sherman tree in Sequoia National Park (see fig. 1.1 in chapter 1), lost a branch in 1978. This branch, more than 6 feet in diameter and 150 feet long, was larger than almost every tree that now grows east of the Mississippi River.[3]

Many species of conifers live in the coastal forests of the Pacific Northwest. The most spectacular of these is the coast redwood, the species that contains the world's tallest trees, many of them exceeding 350 feet in height. The coast redwoods live only in northern California and southern Oregon. Along the coast of Washington, many other conifers, such as the Port Orford cedar dominate the woods. These forests experience relatively mild winters, and abundant fog.

The giant sequoia forests and the Pacific Northwest woodlands are remnants of forests that have survived in little pockets of moist climate in western North America. Many more species of conifers have persisted in Asia under conditions similar to those in which deciduous forests live. In China, the dawn redwood (*Metasequoia glyptostroboides*) has survived essentially unchanged for 30 million years. Some Asian coniferous forests, such as the sugi (*Cryptomeria japonica*) and hinoki (*Chamaecyparis obtusa*) woods of Japan, are remnants of forests that have vanished from North America. Many more temperate conifers persist in Asia than in North America, perhaps because Asian climate has changed less drastically during the past 30 million years than that of North America. As in the deciduous forest, the trees of the temperate coniferous forests create shady, moist conditions in which many other plants and animals live.

Tropical Forests

Finally, as you travel southward from deciduous forests, you enter tropical forests—the warmest and wettest habitats on earth. Their main distinguishing characteristic is the lack of winter. In rain forests, where the rain falls all year long, the trees do not drop their leaves. Some other tropical forests have a dry season, during which the trees do lose their leaves. In North America, the most extensive tropical forests are in the Yucatán

Peninsula of Mexico. The only tropical forests in the United States are in Hawaii, and their trees are much shorter than those of the tropical forests along the equator. The trees found on the hummocks above the marshes of the Everglades in Florida are subtropical, meaning that they are almost but not quite warm and wet enough to be tropical forests.

Tropical forests, because they have the warmest and wettest growing conditions, also have the most plant (and animal) species. A typical tropical rain forest in Peru may contain as many tree species as all of North America. The main struggle of the trees is not against the weather but against one another and against insects that remain voracious all year long, rather than having their populations reduced every winter. Most other tropical plant species are epiphytes, which live in the branches of the trees. They do not take moisture and nutrients from the tree branches but simply use the trees as a place to grow. The abundant rainfall allows the epiphytes to absorb all the water they need from the precipitation that falls on their roots. Therefore, most of the organisms in a tropical forest live up in the treetops, far above the forest floor. If you walk through a tropical forest at ground level, you will miss most of the ecological action. Many tropical biologists now conduct research in the treetops, using balloons, ropes, or aerial walkways.[4]

Although tropical forests have abundant warmth and rain, their soils are usually very poor. Tens of thousands of years of warm rain have leached away the nutrients of the soil. Exceptions include the tropical forests near volcanoes (such as the western edge of the Amazon rain forest, and the Philippines), and along river floodplains. The abundant growth of the plants is not attributable to rich soils but to the tightness with which the plants hold on to the nutrients. The nutrients from leaf litter have no chance to enter the soil but rather are absorbed by tree roots directly from the litter layer. The tropical forest, therefore, can be said to sit on top of the soil, rather than grow in it. The fact that tropical plants create, rather than live in, tropical forests is vividly illustrated by what happens when one is cut down. With the plants gone, all that is left is the poor soil, which supports only a very scanty growth of grass. I discuss reasons for, and consequences of, tropical deforestation in the next chapter.

Western Coniferous Forests

Even under conditions of similar temperature, western North America has very different vegetation than does the eastern part of the continent—primarily because the west is drier. The first example of this is the western coniferous forest. As we travel from the deciduous forests to the western mountain slopes, we encounter woodlands that consist of conifers (table 8.4). The *montane* (mountain) forests of the Rockies and the Sierra Nevada have temperatures similar to those of the eastern deciduous forests, but they are drier. There simply is not enough summer rain in these places for deciduous forests. The lower availability of moisture means that less photosynthesis occurs, and the leaves cannot pay for themselves with a single year's growth.

These montane forests also differ from another in terms of moisture. The wetter ones may contain Douglas firs and white firs. In the mountains of California the montane forests intergrade into forests of giant sequoia, sugar pine, and incense cedar (see above). Moderately dry montane forests are often dominated by ponderosa pines, perhaps the most abundant tree species in the United States. The driest coniferous forests, bordering on grasslands and deserts, are the piñon-juniper forests. Piñon pines prefer wetter zones, and junipers drier zones, within these forested areas. The piñon-juniper woodlands and the Cross Timbers deciduous forests intergrade into shortgrass prairies. The border between woodland and grassland is largely determined by fire, a topic I will explore in the next chapter.

Deserts

Deserts are found in the driest (but not always the hottest) regions. They cover a large area of the western United States and dominate its character. Despite what we see from Hollywood and cartoons, deserts are not empty, and they are not all the same. Television gives us images of saguaro cactus and tumbleweeds. However, cactus is found in only certain deserts, and tumbleweeds (as their more proper name, Russian thistle, would suggest) are not native to North America. Although they contain fewer plant and animal species than do most forests, real deserts are much more complex and diverse than are the deserts of the popular imagination.

Within the land we call desert, there is a great variety of plant communities. *Cool deserts* consist mostly of sagebrush in the Great Basin (table 8.5). The Great Basin is the land between the western mountains (the Sierra Nevada and the Cascades) and the Rocky Mountains. This region receives little moisture because both the Sierra and the Rockies shield it from ocean clouds. The winters are cold and usually snowy, but the summers are hot and dry. The Great Basin is filled with many smaller mountain ranges, topped with piñon-juniper woodlands, making this region a complex sandwich of desert and forest layers. Sagebrush is well-adapted to its environmental conditions. It produces relatively large leaves when soils are wet from spring snowmelt. These leaves are efficient at converting sunlight to food, but they use a lot of water. In the summer drought, the sagebrush drops its spring leaves and produces smaller leaves that make less food but lose very little water. Sagebrush territory intergrades into other types of habitat, such as the shortgrass prairie, which exists largely because of fire (see next chapter). Sagebrush does not grow back very well after fires and is distasteful to cattle. Therefore, fire control and overgrazing have both led to an increase in sagebrush territory.

TABLE 8.5. Locations and Dominant Plants of the Deserts.

	EXAMPLES OF IMPORTANT SPECIES	WHERE TO FIND THEM[a]
Chihuahuan	Cholla (*Opuntia* spp.)	White Sands NM
	Mesquite (*Prosopis glandulosa*)	Big Bend NP
Great Basin	Sagebrush (*Artemisia tridentata*)	Great Basin NP
Mojave	Bursage (*Ambrosia dumosa*)	Joshua Tree NP
	Creosote (*Larrea tridentata*)	Death Valley NP
Sonoran	Catclaw (*Acacia greggii*)	Organ Pipe Cactus NM
		Anza-Borrego SP (Calif.)
	Saguaro (*Carnegiea gigantea*)	Saguaro NP

NOTE: NM = National Monument; NP = National Park; SP = State Park.

[a]Examples of locations easily accessible to travelers.

South of the cool deserts, the landscape is covered by *warm deserts*. These areas seldom freeze, though snow is not unknown. The small amount of rain arrives mostly during the winter but can also come from unpredictable storms that suddenly fill the dry arroyos. Summers can be extremely hot. Because there is little plant growth and decomposition is slow, desert soils are poor. There are three major regions of warm desert (table 8.5). The first is the Mojave Desert of California. Here, creosote and bursage bushes are regularly spaced, almost like a plantation, because the area between them is filled with roots that soak up the little rain that falls. The second region is the Sonoran Desert of Arizona, which is where you should go if you want to see cactus. Many different kinds of cactus grow here, the most famous of which is the saguaro. Cacti predominate in dry, rocky areas, while on the slopes that receive flows of water during rains you will see abundant bushes as well. These shrubs play a very important role: cactus seedlings die in full sunlight and usually germinate in the light shade of one of these bushes (fig. 8.6). You will also see large agaves, with spine-rimmed succulent leaves. They grow slowly for many years, then produce a huge trunk of flowers, after which they may die. The third region is the Chihuahuan Desert of New Mexico and Texas. Creosote and mesquite bushes dominate this landscape, and yuccas are also frequent. Because of its proximity to the Gulf of Mexico, it receives more rain than the other deserts, allowing grasses to frequently grow between the bushes.

Winter rains stimulate the germination of profuse desert wildflowers in the spring. March is a breathtaking time to visit places such as Death Valley or the Anza-Borrego Desert in California, except during drought years. The wildflowers produce blankets of color. Many of the spring wildflowers are ephemeral annuals, which live their entire life cycle, from seed to leaf to flower to seed, in just a few weeks. The entire plant lives only briefly and may be only an inch high, almost half of it flower (fig. 8.7). Other spring wildflowers are perennials, whose underground parts remain alive for many years but which emerge in rapid aboveground growth in the spring. This is also the most likely time for the cacti to bloom. Ocotillo bushes look like dead sticks for most of the year, but in the brief spring they produce leaves and burst into striking red bloom.

[157]

FIGURE 8.6. In the Sonoran Desert, saguaro and other cacti germinate in the shade of shrubs, such as acacias, then grow above them. Photograph by the author.

A Mosaic of Habitats

The United States is home to a great diversity of climatic conditions, from fiercely cold mountaintops to hot, dry deserts. An equally great diversity of plant communities have adapted to these conditions. Along the eastern side of North America, from north to south, we have traced

FIGURE 8.7. In the Sonoran Desert, some annual plants, such as this *Mimulus bigelovii*, are very small and live only for a month. Photograph by the author.

a line of increasing warmth: the arctic tundra of northern Canada is colder than the boreal forest of southern Canada, which is colder than the deciduous forest of the United States, which is colder than southern and tropical forests. A similar pattern results from a descent of a high mountain in the eastern United States: alpine tundra is colder than subalpine forest, which is colder than deciduous forest. Across North America from east to west we have traced a line of decreasing moisture: deciduous forests of Ohio are wetter than the deserts of Nevada. Between the forest and the desert is the grassland, largely the product of fire. In the western United States, temperature and moisture conditions are mixed together, because the mountains, which are home to coniferous forests, are both cooler and wetter than the surrounding deserts. Therefore plants, responding to the general patterns of temperature and moisture, produce the complex, large-scale pattern of habitats in North America.

Little Worlds: Plants and Local Variation

When we look even more closely, we find that within the scale of a few miles, or even a few yards, drastic differences can occur in temperature, moisture, or soil conditions, resulting in strikingly different types of plants: small-scale *microclimatic* patterns. How small is small? Microclimates, like the larger climatic zones, come in a whole range of sizes: as large as a lake or a mountain, or as small as a rock or a hole in a tree. The microclimate is, in fact, the world that an organism experiences. A microclimate may be strikingly different from the prevailing regional climate for several reasons, of which I will explore three: mountains, bodies of water, and edaphic conditions.

Mountains

Mountains create zones that are colder or drier than the surrounding region and produce rain shadows. Mountains therefore create a complexity of microhabitats.[5] As mentioned earlier, by traveling up a mountain you may pass through the same zones of plants that you would see as you travel north: for example, from piñon woodland to ponderosa pine forest to subalpine forest to tundra in the western mountains. Mountains interrupt the latitudinal bands of vegetation by creating colder microclimates. Mountains are colder because they protrude into higher, colder atmospheric layers.

North of the Tropic of Cancer, the southern slope of a mountain always receives more direct sunlight than the northern slope, because the sun is always in the south. As a result, the southern slope is usually warmer and drier, and the northern slope is cooler and wetter. Because afternoons are usually warmer than mornings, it may be more accurate to compare warm, dry southwestern slopes with cool, moist northeastern slopes. In many cases, drought-tolerant plants dominate the former while trees grow on the latter (fig. 8.8).

Mountains may also cause *rain shadows*, in which more rain and snow fall on the windward than on the leeward side of the mountains. In most cases, mountain ranges run north and south, causing either the east or west slope to receive more precipitation. Warm, moist air from an ocean blows inland and encounters a mountain range, which forces the air upward. As the air cools, the moisture condenses, forming rain or snow on the windward slope. As the air, now relatively dry, passes over the mountain, it descends, becoming warmer and therefore even drier. The major rain shadow in the United States is the western desert. Immediately east of the Sierra Nevada in California, deserts occur at elevations that on the other side would be covered with montane forests (fig. 8.9) because the mountains block the moisture from the Pacific Ocean. The Cascade Mountains of Washington and Oregon create a rain shadow in which dry grassland, the Palouse, exists just east of the Pacific Northwest rain forests. The Great Basin is in a double rain shadow, shielded from Pacific moisture by the Sierra and Cascades, and from Gulf of Mexico moisture by the Rockies.

Rain shadows can be found in several other parts of the world, for example in England (where more rain falls on the Gulf Stream side of the Windermeres than on the eastern side) and in Madagascar (where the central mountain range blocks Indian Ocean moisture and creates dry scrubland in the west). Japan even has a snow shadow. Winter storms from Siberia dump snow on the northwestern part of the main island of Honshu.

Aquatic Habitats

The presence of water creates a habitat different from the surrounding areas, even in places that receive plenty of rain. This occurs largely

FIGURE 8.8. On the dry, eastern side of the Sierra Nevada in California, trees grow on the relatively cooler, wetter, northeastern slopes, while desert shrubs grow on the southern and western slopes. The larger desert shrubs (Joshua trees) grow along a watercourse that is usually dry but underneath which the soil is moist. Photograph by the author.

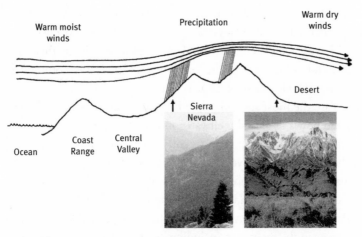

FIGURE 8.9. The Sierra Nevada are the tallest mountains in the continental United States, and they create a rain shadow. The eastern slope of these mountains has desert shrubs at its base at 7,000 feet above sea level. At this elevation on the Pacific slope, there are extensive montane forests. Illustration and photographs by the author.

because plants require special adaptations for survival in continuously flooded conditions. Because roots need oxygen, most plants will die in flooded conditions, as their roots suffocate.

Aquatic habitats can be classified as marshes, swamps, or bogs. Grasses and grasslike plants dominate *marshes*, whereas *swamps* have trees. *Bog* vegetation grows in a floating mat of moss. One of the largest marshes is the Everglades. Where freshwater mingles with saltwater, *salt marshes* may occur, some parts of which are dominated by grasses and grasslike plants, and others by a carpet of *Salicornia* pickleweed. Many marsh plants have internal passageways that allow oxygen from the air to penetrate down inside the underwater stems and roots. Many swamps in the eastern United States are dominated by bald cypress trees (*Taxodium distichum*), so-called because unlike most conifers these cypresses are deciduous. Because trees are unstable in loose mud, bald cypress trunks have wide bases. The roots obtain oxygen by growing "knees," filled with air spaces, above the water level. The water surface may be covered with floating plants such as duckweed. The cypress

[163]

knees and stumps of dead trees form almost the only dry surface on which other species of plants can germinate. Herbs and bushes grow on the knees, which are terrestrial microhabitats within this aquatic microhabitat. Coastal saltwater swamps are typically dominated by mangroves, which are thick, large bushes.

Bogs are the most specialized and unusual aquatic habitats. They are clay-lined bowls of glacial meltwater and are therefore most common in the northern regions that were recently glaciated. Unlike other aquatic microhabitats, bogs receive no influx of mineral nutrients from rivers or streams. The water is therefore very low in nutrients. One kind of plant that thrives in these conditions is sphagnum moss, which forms rafts that float on and may entirely conceal the surface of the water. Larger plants grow in the moss mat, including heath bushes such as cranberries. Not only is the bog nutrient-poor, but the moss releases acid into it, inhibiting the growth of larger plants even further. Moreover, the mat of moss keeps the water beneath it cool. Many bog plants, such as the larch and the cotton sedge *Eriophorum*, are adapted to nutrient-depleted, wet, cool conditions and can be found in the boreal forest or tundra. Indeed the bog could be considered a southerly remnant of the tundra.

The most famous plants of the bog are the carnivorous plants, which trap and digest insects. The leaves of pitcher plants form cylinders into which insects fall. The hairs point downward, making it difficult for the insects to climb out. At the bottom of the cylinder is a pool of digestive juices. The leaves of sundews are covered with sticky hairs that act like flypaper. The leaves curl around and digest the trapped insect. Venus flytraps are found not in true bogs but in wet, acidic soils of pine savannas in the Carolinas. When an insect touches the "trigger" hair, the leaf blade closes around it and digests it. Why do carnivorous plants eat insects? The plants are green and make their own sugar through photosynthesis. Therefore, they do not eat insects in order to obtain food energy. Instead, they use the insects as a source of fertilizer, such as nitrates and phosphates, which are scarce in the acidic moss or soil.

Decomposition is very slow in the cold, acidic waters of a bog. Bogs may therefore hide many ancient mysteries. Human corpses, some thousands of years old, have been found well-preserved in bogs, although the hair has been bleached and the skin literally tanned like leather. Analysis of the corpses of "bog people," particularly in Europe, has helped to reveal facts about the daily lives of people in prehistoric cultures. Therefore bogs, besides being interesting microclimates, act as time capsules that preserve bits of the past.

Deserts also have relatively moist microhabitats. Arroyos, or desert washes, contain plants different from those that dominate the surrounding warm desert. Soil in arroyos is, on the average, wetter for longer periods of time than the surrounding desert. Species of plants can be found here that cannot tolerate cold temperatures, yet require more moisture than the desert usually provides. Many of these plants are remnants from plant groups that covered this region millions of years ago when there was more rainfall and may be members of tropical plant families only rarely represented in the United States. For example, although millions of palm trees have been planted throughout the Southwest, *Washingtonia filifera* palms grow only in a few arroyos. In many desert arroyos, native trees are being displaced by invading salt-cedars introduced from the Middle East.

Edaphic Conditions

Edaphic conditions refer to the material in which a plant grows, usually but not always soil. It can also refer to rocks on which plants grow or the floating mat of moss that covers a bog. Normal, rich soil is a complex structure of many components, a world in itself (chapter 7). Edaphic conditions, such as rocks or unusual soils, can cause an almost entirely different plant community to dominate over a small area, as shown in the following examples.

Consider what happens with rocky, sandy, or shallow soils. Such soils are both drier and poorer in nutrients than normal soils. In Michigan, areas of rich loam soil alternate with poor sandy soils left by glaciers when they melted. Beech-maple forests dominate the rich soils, while jack pines dominate the sandy soils. Jack pines are more commonly found in the poorer soils of the boreal forest. Hemlocks (*Tsuga canadensis*) are often found on poor, rocky soils in the eastern deciduous forest.[6] In parts of Illinois, normal soils also alternate with sand deposits. Black oaks dominate the normal soils, but the sand deposits are so nutrient-poor and dry that no trees at all grow on them. Instead, grasses that are more typical of the western prairies, and even beavertail cacti, grow there. Sand dunes along beaches have different plant species than surrounding habitats. Small cacti grow on granite outcrops in regions of Oklahoma where oaks are the dominant plants. These edaphic conditions, by creating dry microclimates, compress several hundred miles' worth of climatic change into a few yards.

Along the coast of northern California, tall conifers such as redwoods dominate the normal soils. In some small areas, the soils are very shallow and acidic. In these places, rather than tall redwoods, you see a "pygmy forest" of short bishop pines, tiny cypresses, and plants such as rhododendron bushes that are typical of acidic soils. The trees resemble bonsai trees because, just as bonsai trees do, they grow in very poor soils.

Some soils contain toxic minerals that inhibit the growth of most plants. For example, soils derived from serpentine rock contain high levels of chromium and nickel. These rocks resemble those of the earth's mantle rather than the continental crust, and they have been lifted up by the movement of the continental plates. Therefore, serpentine deposits are most common near California's San Andreas Fault. Some species of plants, or varieties of plants within species, are able to grow in these toxic soils. There is a cost to this adaptation, however. These plants grow more slowly in normal soils than do the normal plants. This is what restricts such plants to the patches of toxic soil.

Another unusual soil is *diatomaceous earth*, an almost perfectly white, powdery soil formed from rock that is made of the shells of billions of diatoms (single-celled plantlike organisms) that accumulated on the bottoms of ancient oceans. I undertook my earliest botanical research project on Harris Grade in the Purísima Hills north of Lompoc, California, where large deposits of this white dirt support a forest of bishop pines (*Pinus muricata*). These pines are found nowhere else in the immediate vicinity. The soil factors that may allow this unusual plant community have not been studied very much beyond my amateur undergraduate project.

Rock outcrops can also create microenvironments. A rock outcrop in a forest is a relatively hot, dry niche. A lichen is a mixture of fungi and single-celled algae fused into a single organism. Lichens, which tolerate periods of extreme dryness, often grow on rock surfaces. Meanwhile, plant roots can grow in shallow soil deposits and penetrate cracks. In the ponderosa pine forests of the Black Hills in South Dakota, it is common to see cactus, sumac, and juniper growing on dry rock outcrops. But rock outcrops also offer recesses with a shady, moist microenvironment. Ferns can live on the same rocks that may also support cactus: ferns in the shady recesses, cactus in the dry, exposed spots.

Rock outcrops in the dry grasslands of California provide an example of complex microenvironments. Shallow patches of soil accumulate on the rocks; some very small grasses and wildflowers, such as *Mimulus dudleyi*, are found there but not in surrounding soils. These same rocks create patches of moist soil at their bases owing to runoff during rain. Certain species of wildflowers, such as *Phacelia cicutaria*, grow almost nowhere else in these fields except at the bases of rocks. Moreover, two adjacent rocks may form a shaded, moist microenvironment with yet other plant species.

I grew up within sight of dry hillsides in California, on which grasses and wildflowers dominate the normal soils, while blue oaks (*Quercus douglasii*) grow in soils that have granite boulders (fig. 8.10). Researchers have shown that grasses use water from the soil and can thereby cause oak seedlings to die.[7] Perhaps the oaks grow around boulders because grasses do not grow well there, and the boulders allow water, and tree roots, to penetrate more deeply into the soil. Grazing mammals eat oak leaves and damage seedlings; perhaps the animals do not graze in boulder fields as much as on smooth grassy hillsides.[8] Edaphic conditions (in this case, granite boulders) enhance oak growth, for one or all of the above reasons.

These are just a few examples of microclimates within the larger climatic zones. Because microclimates contain species not found in the surrounding region, their presence increases the overall species diversity

FIGURE 8.10. On these California hillsides, blue oaks (*Quercus douglasii*) grow only where there are boulders. The reasons are unknown. It may be because the boulders allow rain to penetrate more deeply into the soil; it may be that the boulders discourage competition from grasses; or perhaps the boulders provide a refuge from grazing cattle. Photograph by the author.

of the larger plant zone. They are little worlds, with unique climates and unique species, and they make the larger world richer and more interesting. As you travel, watch for these plant patterns on both large and small scales. Indeed, you can read the landscape like a book: large-scale temperature and moisture patterns are the chapters, and microclimates are the paragraphs. And just as plants create the major habitat zones, they also create the minor ones. Without these different plant life zones, the surface of the earth would look about the same everywhere.

The Future of Plant Habitats

The survival of all of these habitats depends on plant responses to temperature and moisture conditions. Human activity, however, is changing both (chapter 3). Overall, the world is becoming warmer, and in some regions rainfall will increase, whereas in others it will decrease. This could spell disaster for the habitats described in this chapter.

Some of the habitats will almost certainly disappear. One example is the tundra. When tundra leaves and stems die, cold temperatures prevent their complete decomposition. Over millions of years, thick layers of organic material have accumulated, in some cases to depths of hundreds of feet, all but the top few inches of which is frozen. Global warming will cause much of the permafrost to thaw, and decomposition will release the carbon into the atmosphere—perhaps as much carbon as human activity is now producing. A warming of the tundra, especially in the vast reaches of northern Alaska, Canada, Europe, and Siberia, will amplify the greenhouse effect. The greenhouse effect will then destroy these same tundras. Longer summers and milder winters will allow the coniferous forests to migrate northward to the Arctic Ocean and up to the tops of the mountains.[9] Low tundra vegetation persists only where it is too cold for trees to cast shade on them. If global change occurred slowly, then perhaps the arctic tundra could establish itself on the now barren islands near the North Pole, but the tundra may vanish before this has a chance to occur. Anyone reading this book fifty years from now may find the description of the tundra to be mainly of historical interest.

The tundra is not the only habitat that will almost certainly vanish as a result of global warming. Warmer and drier weather may kill the temperate coniferous forests that depend on wet and relatively cool conditions. The same is true of the cloud forests on the summits of some tropical mountains. The shrublands of South Africa have a rich assortment of plant species found nowhere else in the world. Stranded on the cape of southern Africa, they will have nowhere to go as climatic conditions change. A large share of world biodiversity consists of species that are found on only one or a few islands. As the climate of these islands changes, or as the islands themselves vanish beneath the ocean waves, their species will disappear.

Other habitats will, it may be supposed, migrate. Boreal and deciduous forests may migrate northward, with the result that the beech-maple forest characteristic of the Ohio River Valley in the United States may be found only in eastern Canada.[10] The problem is that the climate is changing so fast that the migration of plants cannot keep up with it. During the period of rapid climate change, we may find ourselves living amid forests of dead trees, while we are waiting for new kinds of plants to grow up beneath the dead trunks. The forests of southeastern North America may consist largely of dead trees, as a result of drought, during the transition period when grasses and shrubs are beginning to move in and replace them.[11] This is not merely a speculation. Already, large areas of boreal forest are dying as temperatures have risen, and large areas of piñon-juniper woodland are dying as drought has become frequent and severe. In both of these cases, the climate changes have not been severe enough to kill the trees, but rather, they have triggered outbreaks of beetles that have done so.

Large parts of the tropical rain forests will become dry enough that they can support only grassland. However, other parts of the tropical rain forests may become both warmer and wetter, resulting in a set of conditions that exist nowhere on Earth at this time. Many habitats of the future will have combinations of climate variables that are not analogous to anything on the planet today; scientists have referred to this as the "no-analog future."[12] Therefore, the future of plant habitats on Earth may not simply be a different pattern of habitats but disastrous destruction during a transition to an essentially unpredictable future.

chapter nine

PLANTS HEAL THE LANDSCAPE

A people without children would face a hopeless
future; a country without trees is almost
as hopeless.
—THEODORE ROOSEVELT

The Forest *Not* Primeval

Welcome to the tropical rain forest of Central America. Foliage grows luxuriantly everywhere; woody vines bind together the branches of tall trees. In what appears to be an undisturbed, primordial rain forest, a fine mist falls on the leaves and drips from their tips onto the walls of a Mayan ruin. Only six hundred years ago, very little of this rain forest existed. The Mayan building was once part of an extensive civilization, in which three-quarters of the landscape had been transformed into human habitations, centers of community activity, and intensive agricultural production. When, after flourishing for nearly a millennium, the Mayan civilization collapsed, its buildings and walls were reclaimed by the natural rain forest. We think the recovery was complete, but of course we do not know, as the "original" rain forest, from pre-Mayan times in this location, no longer exists for comparison.

In fact, you cannot find an "original" forest anywhere. Or an original grassland or shrubland or desert. All of the natural habitats of the earth, determined by climatic conditions and defined by the plants that both grow in and shape them (chapter 8), have come and gone, and shifted in location, during just the past few thousand years. Sometimes this has been from the waxing and waning of human activity; sometimes it has been from fires or other natural destructive events. But over and above all of these factors, the vegetation has shifted back and forth owing to the advance and retreat of glaciers during and after the most recent ice age. When Henry Wadsworth Longfellow wrote in his epic poem *Evangeline*,

> This is the forest primeval. The murmuring pines and the hemlocks,
> Bearded with moss, and in garments green, indistinct in the twilight,
> Stand like druids of eld . . .

he was describing an Acadian forest that had not been there when glaciers covered the land a few thousand years earlier. The entire surface of the earth is covered with plants that are always shifting locations in response to changes in climate and to disasters and, more recently, to human disturbances. Nevertheless, natural habitats of plants and animals are not fragile crystals that can be easily shattered by one human error or stroke of natural bad luck. They form a web of ecological interactions and consist of a great diversity of species that actually make them strong, not fragile, and able to adjust to fluctuating and sometimes disastrous circumstances. This chapter is their story of dynamic survival and the essential role that plants play in it. Plants not only create habitats; they heal them after disasters.

However, as robust as plants are in healing the disturbances of the landscape, there is only a finite amount that they can do. For millions of years they have successfully reclaimed the living landscape from glaciations, fires, landslides, and volcanic eruptions; for thousands of years, they have responded creatively to human disruptions—the fires we have started, the fields we have farmed, the woodlands we have chopped. As we saw in chapter 1, the recovery has not been complete. The healing of the landscape has sometimes left scar tissue. There is only so much upheaval from which the natural ecological communities can recover. Today, we are challenging plants with an extent of disturbance, and types of disturbance, to which they have never before been subjected. Plants can help to save us from the greenhouse effect (chapter 3), and from droughts and floods and soil erosion (chapters 5 and 7); they will also heal up the injuries we inflict, if we allow them. To an ever increasing extent, we do not.

Disturbance

Strictly speaking, there is no such thing as a forest. "Forest" is the name we give to the set of species of plants and animals and microorganisms that grow together in one location. These species interact with one another, but they are not parts of a forest the way organs or cells are part of a body. If a disturbance such as a fire wipes them away, other species of plants, animals, and microorganisms live in the new environment created by the fire. Because of ecological changes that occur during the subsequent decades, the original set of species may in fact return. But it was the individual plants and animals within each species that returned—not the forest.

Disturbances are of many kinds. We recognize as a disturbance any process that significantly damages the dominant vegetation in a location. If a tree falls in a forest (whether anyone hears it or not), it produces a very small-scale disturbance: in at least the area of forest canopy that the tree once dominated, there is an open space. The many kinds of disturbance can be classified into two categories. First, some disrupt the soil itself, producing a brand-new surface. Examples include volcanic eruptions and rockslides. Second, some disturbances damage the vegetation without damaging the soil. Examples include fires, storms, and floods. These disturbances may leave a layer of soil and may even enhance the soil that is present; they also leave seeds in the soil and perhaps even a layer of low-lying plants over that soil. Notice from these examples, including the fall of the individual tree, that a disturbance is not a negative event from every viewpoint. You could consider a disturbance to be an abrupt arrival of new resources—space, light, soil nutrients—that may greatly benefit the plants that were *not* damaged or killed by the event.

If we hope to preserve a pristine natural environment, at least in part, we must learn to safeguard its processes. We cannot preserve it, timeless and immobile like a painting. We need to allow protected areas to undergo natural changes. Human activities such as farming and the building of roads and towns create immense disturbance. If we properly understand the way that natural groups of plants and animals respond to disturbances, we can fit our activities into the processes of nature, making our intrusions no more difficult for nature to handle than nature's own. Repeatedly, however, we have shown our activities to be not just disturbing but disruptive to nature, by preventing or exceeding its capacity to recover.

Habitats That Need Disturbance

The evening news presents live coverage from California, where wildfires, fanned by hot, dry Santa Ana winds, make the bushes explode into flame in a rapidly advancing front. The reporter's voice waxes frantic. Firefighters work round the clock to stop this natural disaster. We humans have to make the natural world behave itself. More to the point, we do not want the fire to burn the houses of people who built out in the chaparral and who insist on living there no matter what, regardless of fires, earthquakes, and soil that so easily erodes into mudslides. Next, it's live coverage from Wyoming, where fires rush through pine forests during an unusually hot, dry summer.

I think you get the point. Fires and other disasters are not only part of nature. Some habitats actually *depend on* these disturbances. The chaparral of southern California, for example, *would not even exist* if it were not for periodic fires. If people insist on building their homes in a shrubland that experiences an inevitable cycle of fires, they need to prepare for the inevitability of fire. Some natural habitats, like the pine forest in Wyoming, may not rely on fires, but they have been strongly influenced by them throughout their histories. To preserve these habitats, we cannot simply put a fence around them and keep disasters from happening. If we did this, these habitats would in fact disappear!

In the previous chapter, we explored the natural habitats that were created by plants that respond to large-scale and micro-scale conditions of temperature and moisture. We now explore natural habitats that are created by plants that respond to natural disturbances.

Habitats That Depend on Fire

The most common examples of habitats that simply would not exist without fires are the prairies or grasslands of the world. I will use the North American prairie as an illustration, although examples can be found in most other continents (table 9.1). An ocean of grassland, the *tallgrass prairie*, once covered major portions of Minnesota, Iowa, Illinois, the eastern portions of North and South Dakota, Nebraska, Kansas, and extended down into Oklahoma and Texas. The tallgrass prairie received enough rainfall so that forest trees could have easily grown there. Nevertheless, they did not, because frequent fires killed the small trees—but not the grasses, which have underground buds.[1] In contrast, shortgrass (semiarid) prairies such as the western prairies of Colorado, Wyoming, and the western Dakotas, do not have enough moisture for trees to grow. They do, however, have enough moisture to support shrubs such as sagebrush. Yet sagebrush does not dominate these grasslands, largely because of fire. Fire kills sagebrush and other shrubs but not the grasses. Fire is not the only factor, but it is one of the major ones, that cause grasslands rather than shrublands to grow in large regions of the high plains.[2] The steppes of central Asia are similar, although they have different plant species. Most of the plant

TABLE 9.1. Examples of Plant Species That Depend on a Fire Cycle.

	SCIENTIFIC NAME	COMMON NAME
Tallgrass prairie	*Andropogon gerardi*	Big bluestem
	Echinacea purpurea	Purple echinacea
	Sorghastrum nutans	Indian grass
Chaparral shrubs	*Arctostaphylos* spp.	Manzanita
	Ceanothus megacarpus	Large-fruited ceanothus
	Coleogyne ramossisima	Blackbush
Forest trees	*Pinus coulteri*	Digger pine
	Pinus muricata	Bishop pine
	Pinus radiata	Monterey pine

material in these grasslands consists of grasses, but most of the plant species are not. The majority of the species are forbs that usually have broad leaves and often colorful flowers that can make grasslands burn not with fire but with color.

Fire does not usually kill the grasses and forbs; it actually benefits them. The plants die back to the ground each winter, then sprout in the spring. After a couple of years, dead stems and leaves have built up a thick layer that not only shades the ground but also ties up the nutrients. Fire transforms dead litter, with trapped minerals, into fertilizing ashes. The sun penetrates to the soil surface, now dark with ash, and warms it more quickly in the spring. Grasses and forbs grow profusely after a fire, if it occurs during early spring or late fall. Fires are unlikely to occur at other times, when the grassland has wet, green foliage.[3]

Prairies, especially those with tall grass, have deep, rich soil. Nearly all of the North American tallgrass prairie has been converted to prime agricultural land. Very few of these grasslands now remain, most of them along railroad tracks or cemeteries that were established prior to agriculture. Preserving these prairies is therefore very important. Because it keeps the trees and shrubs out, fire is a necessary management tool for maintaining prairies. Fall and spring are the best times for controlled burns, as the prairies are not filled with green blades of grass. But these are also the times when the wind can be strongest. There was both a sense of joy a feeling of tension during the prairie burn in which I participated—joy, for maintaining one of the few remaining prairies, and the fear that the flames might escape into adjoining private property.

Another habitat that depends on fire is the chaparral.[4] Found mostly in California, chaparral habitats consist of fire-adapted shrubs (table 9.1). After fire, the shrubs resprout either from their roots or from seeds. If several decades pass without fire, the chaparral shrubs have more dead than living stems, and the litter layer beneath them is deep and crisp in the hot, dry summers.[5] In the chaparrals in which I have conducted research, dead stems formed a nearly impenetrable thicket (in which rattlesnakes hid) and I could smell volatile combustible chemicals produced by the shrubs themselves. Fire is inevitable. After the fire, not only do the shrubs grow back, but many wildflowers as well, whose seeds have lain dormant in the soil since the previous chaparral fire.[6]

There are certain pine forests in coastal California that also rely on a fire cycle (table 9.1). They are called closed-cone pines because their cones do not open when mature. The seeds remain sealed inside until a fire bakes them open. Then, after the fire, the seeds fall on the ground and germinate into a thick regrowth of pine seedlings (fig. 9.1). Ever since my first undergraduate research project, I have been visiting the forest of bishop pines (*Pinus muricata*) in the Purísima Hills near Lompoc, California. A fire destroyed the forest in 1995. Before the end of the summer, the ground underneath standing black trunks was covered with almost nothing other than pine seedlings. Beneficial fungi helped the pines grow rapidly in soil that was fertilized by the ashes.[7] By 2002, some of the new pine trees were already producing cones.

The foregoing examples are of ecological communities that utterly depend on fires. Most ecological communities do not require periodic burning for their survival, but they still may benefit from them. Most habitats would benefit from occasional fires that clear away old wood and create sunny openings for new plant growth.

Some ecological communities benefit more than others from fire. In particular, many pine forests the in western mountains of North America profit from occasional fires. Young lodgepole pines (*Pinus contorta*) often have cones that open without fire, whereas older ones often have cones that will not open unless they are burned. The populations of pitch pines (*P. rigida*) of the pine barrens in New Jersey also consist of a mixture of closed-cone and open-cone trees. The cones of ponderosa pines (*P. ponderosa*) in western North America do not need fire to open, nor do their seeds need to grow on burned soil. However, fires clear out openings in which pine seedlings can grow much more readily than in the shade of trees in an undisturbed forest.

Fire suppression is not entirely successful at achieving even the goal of protecting human property. When fires do not burn in western forests, or chaparral, dead timber builds up. When there are no small fires, the risk of large fires increases. When fires begin as a result of lightning or arson, they are much larger and more damaging than they

FIGURE 9.1. A dense stand of bishop pine (*Pinus muricata*) saplings grows two years after a fire destroyed a forest of bishop pines in the Purísima Hills of coastal California. The pattern of ecological succession in this forest is simple: pines replace pines after disturbance. Photograph by the author.

would otherwise have been. Forest and shrubland managers often clear some areas of forest as an artificial substitute for fires. They also use controlled burns, small fires deliberately set, to reduce the danger of larger fires. Sometimes, albeit rarely, these fires get out of control.

Habitats Affected by Other Kinds of Disturbance

Fire is the most common but by no means the only type of disturbance that determines the species composition of wild habitats. Riparian (riverside) forests consist of tree species such as sycamores, cottonwoods, box elders, and willows, whose seedlings begin their growth in sunny, wet conditions. These conditions are created by floods that wash away the overstory of trees in part of the floodplain and open up a sunny space in which the seedlings can grow. The seedlings do not grow well in shade and therefore do not establish very well in places that have not flooded. Even disturbances caused by animals, usually with human influence, can affect natural habitats. At the border of western grassland and desert, for example, the cacti and desert wildflowers grow best in areas that have been disturbed, in this case by erosion, either natural or from overgrazing.

Humans have spent considerable money and energy to bring natural disturbances under control. For example, we seldom permit rivers and streams to flood, although sometimes they overwhelm our ability to harness them. Humans have worked particularly hard to control fire. We usually try to suppress fires completely, because almost every place is densely populated with houses and almost any fire could damage somebody's property. In some areas of the western United States, ponderosa pine forests are spreading and grasslands retreating because of fire suppression. This has occurred because fires, although they benefit pine forests, benefit grasslands even more. In other areas, sagebrush is overtaking areas that once had a grassland cover, because fires no longer kill the sagebrush, but also because the grass is frequently overgrazed.

Ecological Succession

Plants grow in recently disturbed areas, but they do so in a succession of ecological stages, one after another, until eventually the habitat consists of a group of species that is relatively stable over a long term. This stable set of species is usually the one that characterizes a climate zone—thus in the deciduous forest region, a fire or storm creates an opening, in which, after many years, a deciduous forest again grows. Such a group of species is called the "climax," as if it is the culmination of a story, or a "mature" forest, as if it is an organism growing up. These terms are metaphorical, but still useful. We may think of this process as "recovery" or "healing," because in many cases bare soil is covered more and more completely by plants.

There are as many different stories of ecological succession as there are habitats and disturbances. We can, however, roughly classify them into two categories: (1) *primary succession* occurs when plants grow on a newly created surface; and (2) *secondary succession* occurs when plants grow on or from soil that has been exposed. Notice that these two kinds of succession correspond to the two categories of disturbance mentioned previously—those that destroy the soil of a previous habitat, and those that leave the soil largely intact. In a location in which primary succession occurs, the soil must build up before the climax vegetation can return. Secondary succession has a head start, so to speak, because the soil is already there, and usually already has seeds in it.

Primary Succession

Most of the people who visit the Indiana Dunes National Lakeshore along the southern shore of Lake Michigan, within view of Chicago, do not realize that ecological succession is occurring right before their eyes, or that this very spot was where the process of ecological succession was first studied by botanists.[8] The newest of the Lake Michigan dunes in North America, the ones most recently deposited by wave action, are next to the shore; and the older dunes are found further inland. To walk back from the shore, therefore, is to walk forward in successional time.

Sand is not soil. Plants cannot grow on dry sand. In order for anything to grow on sand dunes, there must be water. The sand dunes in Death Valley will probably remain barren. But those near Lake Michigan receive rain, and the water table is not far below the surface. The Lake Michigan dunes are, however, hazardous places for plants to live. Sand holds very little water. Even after a rain, it dries quickly. Sand also holds very few of the mineral nutrients that are essential to plant growth. And it is unstable. A seedling that begins to grow may soon find its shoot buried, or its roots exposed, by the shifting sand. And when the wind picks up, it can literally sandblast the little plant. Finally, sand can be very hot, as any barefoot person knows. The seedling would receive heat from the sun above and reflected warmth from the sand below as well.

How could a little seedling begin to grow in a sand dune? As a matter of fact, few ever do. Seedlings of the sea rocket (*Cakile maritima*) can grow right next to the water, where the sand is wet, cool, and somewhat stable, but seedlings seldom sprout in the dry dunes themselves. Instead, the dunes are invaded by underground stems of grasses that specialize in living on dunes. These pioneer stems can survive the harsh conditions largely because they receive food and water from the home base of the grass clump.

Here is the crucial step in sand dune succession. The underground stems of grasses hold the sand down and contribute organic matter (dead leaves) to it (table 9.2). This organic matter begins to transform the sand into something more similar to soil, allowing the sand to hold more water and mineral nutrients. It also glues the sand grains together. The stems and leaves slow down the wind right next to the sand's surface. The cohesion of the organic matter and sand, and the reduced wind, keep the sand from blowing away. The stems and leaves cast shade, and leaves transpire, cooling the sand. The grasses, therefore, counteract all of the factors that make sand a bad place for plants to grow. Bare sand is therefore replaced by grass.

Shrubs can then grow within the grass clumps. They stabilize and enrich the sand even more. Later, pines grow from amidst the shrubs.

TABLE 9.2. Some Successional Plant Species on Lake Michigan Dunes.

	SCIENTIFIC NAME	COMMON NAME
Grasses and wildflowers that stabilize the dunes	*Ammophila breviligulata*	American beach grass
	Calamovilfa longifolia	Prairie sandreed
	Lithospermum caroliniense	Puccoon
	Schizachyrium scoparium	Little bluestem
Shrubs that grow from grass-stabilized dunes	*Arctostaphylos uva-ursi*	Bearberry
	Juniperus communis	Common juniper
	Prunus pumila	Sand cherry
Early trees	*Pinus resinosa*	Red pine
	Pinus strobus	White pine
Late trees	*Acer rubrum*	Red maple
	Quercus rubra	Red oak
	Tilia americana	Basswood

Pine seedlings cannot grow in the shade of their parents. Seedlings of many hardwood trees, however, can grow in shade. From underneath the pines, trees such as oak and basswood grow, eventually producing a forest. In the shade of these trees, understory shrubs and small trees such as sassafras can become established. As you walk back from the shore, you pass through zones of grass, shrub, pine, and hardwood; that is a cross-section through time. The development of a forest on the dunes is not a simple, orderly process. Storms can rip out areas that the plants have stabilized, and the sequence of events differs greatly from one location to the next, determined largely by which seeds—of shrubs, pines, and hardwoods—happen to land in which spots.[9] Nevertheless, in general, the succession of vegetation causes, and is caused by, the development of soil (fig. 9.2).

Plants build up the soil in other examples of primary succession also. Mosses grow on rocks, secreting acid that begins to break the rock into gravel. The gravel, dust that blows in, and dead mosses begin to create a soil in which small herbaceous plants can live. Eventually trees may grow in cracks in the boulders; the growth of the trunks cracks the rocks even further.

Mount St. Helens in Washington State erupted in 1980, producing a nearly 37–cubic mile landslide of rock and hot gas that killed nearly everything in its path. In the quarter century since that time, primary succession

Years	0	100	200	400

LOW	Ability of soil to hold moisture	HIGHER
LOW	Nutrients in soil	HIGHER
LOW	Organic matter in soil	HIGHER
SUNNY	Light intensity	SHADY
HIGH	Wind and soil instability	LOW

FIGURE 9.2. Primary succession occurs on the sand dunes of Lake Michigan as plants transform the sand into soil. Illustration by Gleny Beach and the author.

has occurred. Rather than being an orderly process, the region's recovery from the volcanic devastation has taken unexpected turns. The early-successional pioneers have often been the plant species that happened to get lucky and survive in a few spots; thus the course of succession has been a little different in each location. Nevertheless, as in the example of the Michigan dunes, plants have helped to build up the soil, especially those with nitrogen-fixing bacteria in their roots: lupines (*Lupinus lepidus*), with *Rhizobium* bacteria, and alders (*Alnus viridis*) with *Frankia* bacteria.[10]

Secondary Succession

In parts of the eastern deciduous forest, you may walk through what appears to be an intact woodland, only to come in contact with a stone fence, built by farmers near the time of the Revolutionary War to divide their farmlands from those of their neighbors (fig. 9.3). When railroads

FIGURE 9.3. A stone fence once separated what were farm fields in colonial New York; after the 1830s, however, these farms were abandoned, and the forests grew back. Photograph by the author.

allowed farm produce to be shipped east from the newly settled Ohio Valley in the early nineteenth century, many eastern farms were abandoned. Here is what happens to abandoned "old fields" in the eastern deciduous forest zone (fig. 9.4).[11]

[184]

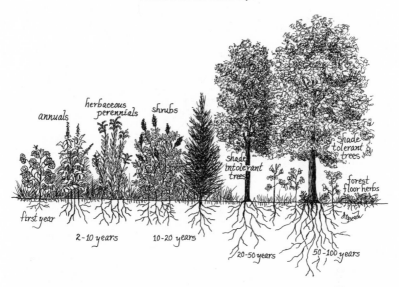

FIGURE 9.4. Secondary succession is a process of shading: shade-tolerant trees overgrow the shade-intolerant trees, which overgrow the perennial shrubs and weeds, which overgrow the annual weeds. Illustration by Gleny Beach.

The first year after abandonment, there is a profuse growth of plants, because the soil is already there and does not need to be built up. The largest plants, however, are annuals such as ragweed, which grow rapidly. At the end of the first growing season they put everything they have into a burst of seed production. Some of the seeds of these annuals may have blown in with the wind, but many were already present in the soil, because these are the same species that grew as weeds amid the crops (table 9.3). After a forest fire, annual weeds may also sprout, some of them from seeds that might have fallen into the soil hundreds of years earlier, during a previous disturbance! Especially when a forest is destroyed, the seeds that sprout from the soil can be of very different species from those that dominated the forest.[12]

Meanwhile, down in the shade of these annual weeds are tiny seedlings of perennial weeds such as goldenrod. You would never know this unless you crawled down under the big plants, as I did during part of my thesis research, to look for them. Perennial weeds grow more

[185]

TABLE 9.3. Examples of Plant Species at Different Stages of Ecological Succession in Eastern North America.

STAGE	SCIENTIFIC NAME	COMMON NAME
Annuals	*Abutilon theophrasti*	Velvetleaf
	Amaranthus retroflexus	Pigweed
	Chenopodium album	Lamb's-quarters
	Conyza canadensis	Canada fleabane
Herbaceous	*Aster pilosus*	Hairy aster
perennials	*Solidago canadensis*	Canada goldenrod
Shade-intolerant	*Acer saccharinum*	Silver maple[a]
trees	*Diospyros virginiana*	Persimmon
	Gleditsia triacanthos	Honey locust
	Platanus occidentalis	Sycamore[a]
	Populus deltoides	Cottonwood[a]
	Prunus serotina	Black cherry
Shade-tolerant	*Acer saccharum*	Sugar maple
trees	*Fagus grandifolia*	Beech
	Quercus velutina	Black oak
Forest floor	*Osmorhiza longistylis*	Sweet cicely
herbs	*Polygonatum biflorum*	Solomon's seal
	Sanicula gregaria	Snakeroot

[a]Primarily in wet areas.

slowly than the annuals even under ideal conditions, because annuals use all of their photosynthetic food for their current growth, whereas perennial weeds store much of their food in underground stems and roots. Moreover, the annual weeds are shading the perennials and taking most of the water and nutrients from the soil. The perennial weeds remain small and do not produce seeds during their first year.

When the second growing season begins, all the plants sprout from the ground. The annual species must sprout from seeds: thus they are small. The perennial plants, however, sprout from underground reserves of food, which usually exceed the amount of food stored in a seed. As a result, they grow faster than the annuals. Starting in the second year the perennial weeds begin to shade out the annual weeds, a process completed a few years later.

Meanwhile, down in the shade of these weeds are tiny seedlings of shrubs such as blackberry and trees such as black cherry. These seedlings may have even been present during the first year, but they grew very little. Once again, you have to look closely to see them. These shrubs and tree seedlings grow even more slowly than the perennial weeds. When later growing seasons begin, all the plants sprout—the weeds from the ground, but the bushes and trees from their above-ground buds. Therefore, from the very beginning of the season, the leaves of the bushes and trees are higher than those of the weeds. The bushes and trees begin to dominate after the fifth to fifteenth years. The trees are, at first, scattered individuals, but later they form a canopy.

These trees are shade-intolerant species (such as black cherry and locust) whose seedlings require bright light to grow. As these trees reach reproductive age, they release seeds. If the seeds land in the shade underneath the parent trees, the seedlings do not thrive. But there are two groups of seedlings that do grow well beneath the shade-intolerant trees: herbaceous plants such as wild ginger and mayapple that cannot tolerate the heat and dryness of full sunlight and seedlings of shade-tolerant tree species such as maple. These seedlings may also have even been present the first year, but their growth was insignificant. Tree species differ in the ability of their seedlings to tolerate shade; many, like oaks, are intermediate in this respect.

The shade-tolerant trees establish themselves in the shadow of the intolerant trees. When the intolerant trees die, the tolerant trees become the new dominant species. Their seedlings can survive in the shade cast by the parents. Forest floor herbaceous plants, such as wild ginger and mayapple, now have the shade that they need to flourish.

During secondary succession, plants that grow later in the process shade out the earlier plants. The perennial weeds shade out the annual weeds; the shade-intolerant trees shade out the perennial weeds; the shade-intolerant trees put themselves out of business by creating shade their own seedlings cannot tolerate; the shade-tolerant trees grow up underneath them and replace them; and finally, the shade-tolerant trees, which can replace themselves, form a climax. But this is not the only way in which plants alter their environment during succession. As this progression occurs, the plants may help to build up the soil. Also, the pioneer plants create moist conditions that help the later plants to grow. The plants smooth out the variability in light, temperature, and moisture, transforming the sometimes wild fluctuations of old fields into the comparatively mild forest floor environment.[13] The result is a complex process in which the earlier plants may either help or inhibit those that assume their dominance later.[14] As a result of these developments, even though the luck of the draw may determine which seeds are present at the beginning of succession, the result is eventually a habitat that is characteristic of the region—in this example, a deciduous forest.[15]

What does this process mean for us? First, forests are almost always undergoing one or another part of the cycle. Natural disturbances continually create openings in the woods, and most forests are patchworks of different stages of succession. Almost all of the eastern deciduous forest of North America was cleared for cultivation and lumber, and almost all of the forest we have today, with the exception of a few remnants, has grown by succession only during the past century or two (chapter 1).

Second, disturbances and succession are actually good for the plants and animals of a region. Without disturbances, old dead wood builds up, and the forest begins to resemble the fictitious Mirkwood of J.R.R. Tolkein's Hobbit novels. The diversity of species and the amount of living biomass actually begins to decline in the absence of disturbances.[16] We can hardly know what such a forest would look like, because disturbances come sooner or later to each one. Think, furthermore, of all the species of plants that simply could not grow in a forest but must live in the patches, and edges of forests, created by disturbance. Disturbance, so long as it destroys only portions of a forest, actually enhances the diversity of plant species. Animal species diversity is similarly enhanced; deer and many birds prefer the rich foraging to be found at a forest's edge over the shaded depths.

After the Ice Age

The biggest recent disturbance of all, in North America, was the ice age. Every hundred thousand years for about the past two million years there has been a cycle of glacial advance and retreat in the northern hemisphere, each one producing an ice age. The most recent glaciation ended about fourteen thousand years ago. The earth was quite a different place at the height of the previous glaciation. Because so much water was frozen in the glaciers, sea level was almost 300 feet lower than it is today. Much of the continental shelf, which today is under shallow water, was dry land. Some "fishing banks" of shallow water in the oceans were exposed islands or peninsulas during the glaciation. Most of the same vegetation zones that exist today were present, including tropical rain forests, seasonal forests, and savannas. However, because the glaciers came so far southward, the northern vegetation zones were thinner, and compressed toward the equator (fig. 9.5).[17] The tropical areas, though slightly cooler, were still warm. However, because so much water was

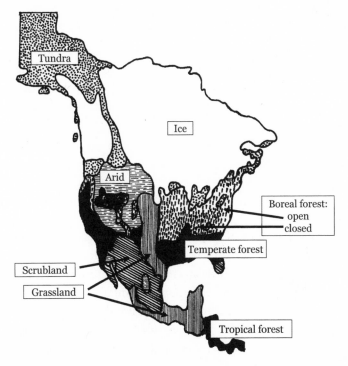

FIGURE 9.5. During the most recent ice age, vegetation zones were compressed southward by the cold weather and the ice sheets. Illustration by the author, based on note 3.

frozen into glaciers, the weather was drier. Tropical rain forests were smaller, and tropical savannas and grasslands more extensive.

As the glaciers began to melt, the land underneath was exposed. Virtually all life was extinguished under the glaciers. Some of the land was scraped flat and clean, while in other places the scrapings were piled up into moraines. Melting glaciers poured water into huge rivers. Immense lakes formed behind ice plugs, and when the plugs gave way, massive floods occurred. Into this violent landscape, tundra vegetation returned from the south, as seeds were carried by wind and animals. Boreal forest trees also migrated northward. Tundra and boreal trees also migrated up mountains, where they remain, stranded, as alpine tundra and subalpine forests (chapter 8). Tundra and forests did not migrate as a unit, but each species (actually, each individual seed) traveled separately. Thus, after the ice age, there were some combinations of species (for example, the spruce savanna) that are no longer found together.[18]

At the end of the ice age, the weather got even warmer and drier in some areas than it is today. It was at this time, during a postglacial warm period, that the great American grassland was established about seven thousand years ago. As the weather became a little cooler and wetter about five thousand years ago, the prairie remained in place because of the fire cycle (see above).

Human Disturbances

The coming and going of the ice ages shows that vegetation can recover from extensive disturbance. It would seem that turmoil caused by humans could not possibly be any more severe than this, so there is little cause to worry about our effects on the natural environment. Why be concerned about clear-cutting a mountainside, when a glacier has already done this, and the forest eventually recovered? The answer is that disturbances caused by humans are different from and often more severe than natural ones.[19]

Human disturbances are more extensive and severe. Fires, floods, storms, landslides, and other disturbances have destroyed forests not very differently from the way humans do so. But humans destroy whole tracts of forest, which amounts to much more extensive damage than any natural catastrophe except for an ice age. No other natural disturbance has so completely cut down the deciduous forest, for example. Human disruptions can be more severe as well. Bulldozers can produce a completely bare soil, which natural disturbances almost never do.

Human disturbances occur much more rapidly. In a century, we have destroyed as much vegetation in North America as the glaciers did in more than a thousand years. Nature can handle disturbances, and even depends on them, as long as they do not occur too quickly. We cause disturbances roughly ten times as fast as they occur naturally. Because of the carbon dioxide we are putting into the air (chapter 3), global temperature is rising at least ten times as rapidly as it did while the glaciers were melting. Global warming is nothing new to this planet; global warming at this rate is unprecedented. You can catch a bullet moving at ten miles an hour, but not one that moves at a hundred miles an hour.

Humans have created impediments to migration and succession. As natural global warming occurred while the glaciers retreated, there was plenty of room for seeds to be carried and for animals to migrate northward. They cannot do so in response to the current global warming caused by human activity. There are too many cities, highways, and farms in the way. We confine wild species to tiny nature preserves and insist that they stay there. When the climate changes, as it inevitably will, wild plants and animals will not survive in the little enclaves to which we have relegated them, and they will be unable to leave. Conservation scientists are making plans for corridors of access from one national park or forest to another, but such plans are too small to make much difference in the greenhouse world that will soon arrive.

Many human disturbances are unnatural. When humans chop down a forest or plow a grassland, our effects may be more severe but are not strikingly different from natural processes. However, our toxic wastes are unlike anything found in nature except for occasional volcanic discharges. When air pollution causes acid fog that kills mountain forests, the forests are helpless because natural selection has never before operated to cause the evolution of acid-resistant strains. Acid resistance cannot evolve now because the air pollution has occurred too rapidly. In a hundred thousand years, plant populations may evolve resistance to air pollution—but not in a hundred years.

One example of human disturbance exceeding the capacity of ecological succession to respond is the current destruction of tropical rain forests. Small disruptions, caused for example by storms, occur frequently in the rain forest, and it grows back quickly from them. In tropical forests, as in other forests, many herbaceous and tree species depend on these little disturbances to create a place for them to live.[20] Farmers who practice "slash-and-burn" agriculture cut down small rain forest clearings and burn the plants, grow crops for a while, and move on when the soil is exhausted (which occurs rapidly in these poor soils). The forests grow back when leaf litter and seeds fall from surrounding trees, and birds drop seeds as they fly from trees on one side to the other. Rain forests have survived, even benefited from, these disturbances for millennia. In fact, it is estimated that much of the Amazon rain forest was cultivated prior to the introduction of smallpox and other diseases by the Spaniards about 1500, which decimated Native American populations (chapter 1).

However, more recently, huge tracts of rain forest have been destroyed for large-scale agriculture, ranching, and other purposes. These clearings are too large for leaf litter and seeds to fill them or for forest birds to fly across them. The big area in the full tropical sunlight becomes too hot and dry for most plants to survive, unlike the small clearing that is shaded by the adjacent trees. In some cases, these large clearings are not able to undergo succession back into tropical rain forests except, perhaps, over periods of time far longer than is normal for tropical succession. For all practical purposes, in human lifetimes, these forests have been destroyed. If a few large trees happen to survive, or if people carefully plant and maintain new trees in these openings, however, the forest may grow back. People must water the replanted trees and keep the voracious leaf-cutter ants and vines from overrunning them. Once the trees are large enough to survive, they become nuclei that allow rain forest succession to recommence: their leaves and seeds drop in an expanding circle around them, and they provide a haven for birds to stop when flying across the opening. Although it is not necessary for humans to replant the entire opening, they must replant much of it to allow the opening to be refilled by ecological succession.[21] Such intense labor, almost exclusively volunteer, is not available for most of the extensive tracts of rain forest destruction. Fortunately, many of the disturbances in the Amazon basin are small enough that rain forests and other tropical woodlands can reclaim them: 32 percent of the deforested landscape in the Amazon basin is undergoing succession.[22]

The Invaders

One example of a disruption caused by humans is that we have transported and introduced exotic species.[23] By "exotic" I do not mean a tropical paradise; the word means "from the outside." When a species is introduced into a new location, it will most likely find the new conditions unsuitable. Sometimes, however, it finds the new conditions ideal. It may then experience a population explosion, because its natural competitors, herbivores, predators, and diseases have been left behind. This process occurs naturally. For example, when placental mammals migrated over the Panama land bridge into South America about a million years ago, they seem to have had population explosions that helped send many New World marsupial species into extinction.

But humans have accelerated the process of exotic species introductions. When humans move to new locations, we like to take our old buddies with us, and some of them even hitch a ride without our knowing it. We have taken livestock and pet animals and crop and horticultural plants with us. Animal pests, such as rats, and weeds that grow in cultivated areas, have sneaked along as well. Feral or escaped pets (such as cats in Australia) and livestock (such as pigs in North America) have produced large and destructive populations, especially on islands, such as the Hawaiian Islands, where pigs and goats have destroyed large areas of native forest.

There are many examples of exotic plants that have experienced population explosions. Crops are now planted worldwide in suitable climates. More wheat is grown in North America than in its native Middle East, more potatoes in the northern United States than in their native South America, and almost as much maize in Africa as in its native America. Crops do not usually act as exotic invaders, however, because crop plants are unable to reproduce and spread without cultivation (chapter 10). But weeds, which reproduce and spread on farms, have often been successful invaders. Weed seeds have arrived in new locations through inadvertent mixing with crop seeds or via the soil of potted plants. Although agriculture originated in America independently from the Old World, it began later; there has been more time in Eurasia for the evolution of weeds. This is why most American weeds (for example ragweeds and cockleburs) came from Eurasia.

People have also carried their favorite garden plants with them, and some of these plants have escaped into the wild. Kudzu and purple loosestrife are two examples (see below). A particularly interesting story is velvetleaf (*Abutilon theophrasti*). It was raised in American gardens because its star-shaped immature fruits were used to make decorations in slabs of butter; hence two of its common names, butterprint and stampweed. Its seeds escaped into fields, and today it is one of the major weeds of soybean and cotton fields.

Some plants have had more unusual methods of transport to new locations. Puncture vines (*Tribulus terrestris*) grow prostrate on the ground and their seeds have sharp spikes that make them stick painfully into the paws, hooves, and feet of animals, including humans. They also stick in automobile tires, without necessarily deflating them. Livestock and automobiles have carried puncture vine seeds to new destinations.

Sometimes we deliberately introduce plants into a location to change it, and only later do we regret it. Melaleuca trees (*Melaleuca quinquinervia*), which transpire like crazy, were introduced into the Everglades to suck the water out of them for the development of residential areas. They are doing precisely that. Now that we want to preserve the Everglades instead of destroy it, we find that the melaleuca trees are a major problem. Because each tree produces many thousands of seeds a year, it is almost impossible to eradicate them.

Population explosions of introduced plants can have devastating consequences on the natural vegetation. Purple loosestrife (*Lythrum salicaria*), imported from Europe, produces millions of seeds and displaces native

FIGURE 9.6. Kudzu covers a pickup truck that was parked a little too long in Georgia. Photograph by the author.

plants from wetlands. The native animals are not adapted to use loosestrife as a food source. Dense populations of introduced plants can crowd out almost every other species, as melaleuca trees are doing in the Everglades. Introduced vines can crawl over and literally cover everything in their path (fig. 9.6). Better not leave your truck parked too long in kudzu territory! Kudzu, an ornamental vine that escaped, grows over even very large trees and kills them. Introduced plants can shed leaves so thickly that no other plants can grow underneath them, as casuarina trees are doing in Hawaii. Attempts are being made to control these plants, using everything from the introduction of insects to eat them, to the radical surgery approach of bulldozing them away.

What can we learn from the study of disturbance and succession and plant responses to other disruptions? Primarily, we see that nature is not fragile. It will recover if we give it a chance. Plants will grow back and heal the land, if we let them. However, we are disturbing the earth far too much, too fast, and in ways from which even those superheroes, plants, cannot save us.

chapter ten

HOW AGRICULTURE CHANGED THE WORLD

I wish I was a despot that I might save the noble, the beautiful trees that are daily falling sacrifice to the cupidity of their owners, or the necessity of the poor. . . . The unnecessary felling of a tree, perhaps the growth of centuries, seems to me a crime little short of murder.
—THOMAS JEFFERSON

Plants keep the world, including the human world, alive in many ways, as explained in the preceding chapters. But there is a particularly special way in which plants maintain our existence and support the entire human economy: agriculture. In prehistoric times, when there were only a million people in the world, hunting and gathering was sufficient to feed the human species. But for thousands of years, most humans have been dependent not on wild foods but on food that they raise on farms. Today, with more than six billion people, the survival of most humans would be unthinkable without agriculture.

We tend to think of agriculture as the use of crop plants for human benefit. We think of crop plants as our slaves. But these plants also benefit from agriculture. In fact, crops depend on humans. Agricultural plants have been bred into forms that are unable to survive in the wild. It is just as valid to think that humans serve the interests of the crop plants as it is to think that they serve us. We prepare the ground, plant the seeds, water and fertilize the plants, harvest the seeds, and save some of them for the next year. As science writer Michael Pollan points out, the most successful plants are those that get us to serve them by producing things that we need, or that we simply want. We have turned much of the forest and prairie into an 80 million–acre corn lawn, says

Pollan; by creating cornfields, we have swept away all of the other plants with which corn would otherwise have to compete.[1]

An Agricultural History of the Human Race

Historians have long debated how agriculture began. Some scholars used to believe that agriculture was invented by a brilliant man in a tribal society of hunter-gatherers. Other scholars pointed out that, because women gathered most of the plant materials, agriculture was probably invented by a woman. Both the brilliant-man and the brilliant-woman theories are incorrect, however, because agriculture could not have been invented in a single step by anyone. Agriculture had to evolve, because unmodified wild plants, such as the wild ancestors of wheat, are unsuitable for agriculture, in four ways. First, the seeds of most wild plant species are dormant when they are mature. That is, when planted, the seeds will not germinate. Many wild seeds require a period of exposure to cool, moist conditions before the seed can germinate. If a brilliant man or woman planted a seed from a wild plant, it would not have grown, and he or she would have rightly concluded that agriculture was not a good idea. Second, the seeds of many wild grains shatter (fall off of the stem) as soon as they are mature. Because the whole point of the seed is to grow in a new location, shattering is beneficial to the plant, but it is extremely inconvenient for a human harvester. Third, the seeds of many wild plant species contain toxins. And finally, the seeds of many wild grains are small.

Furthermore, the advantages of primitive agriculture would not have been immediately apparent to intelligent hunter-gatherers. Agriculture requires intense labor. Modern hunter-gatherers often barely eke out an existence in marginal habitats such as the Kalahari Desert or Great Outback; but these are the habitats to which tribes and nations with more advanced tools have driven them. Before agriculture, many tribes lived in rich habitats in which hunting and gathering in many cases provided a comfortable level of existence. For these reasons also, agriculture had to evolve gradually.

Agriculture, therefore, required a gradual origin, through an evolutionary process, which most likely occurred in the following way.[2] Through most of the history of the human species, omnivorous people hunted animals and gathered plants. Because of the toxins and other defenses in the wild plants, humans had to carefully select the plant materials—digging certain roots, eating certain fruits and nuts, and avoiding poisonous species or parts. If the seeds of wild grasses fell off of the stem as soon as they matured, our ancestors would not have been particularly likely to gather them and take them back to camp. Moreover, people preferred to gather large grass seeds, and grass seeds that tasted good. They would sometimes take the food they had gathered with them when they went to a new location; and in the new location, they may have thrown the seeds on the ground, especially their favorite plants from "back home." When they did so, any of the seeds that had complete dormancy would not have grown. The ones that by genetic chance had the least dormancy would have sprouted, however. These gatherers would have gradually selected (within each of their favorite wild species) those breeds that did not shatter, that did not have dormancy, that had large seeds, and that did not have toxins. They would also have selected for a high *harvest index*. Harvest index is the weight of the edible part of the plant divided by the weight of the entire plant. Harvest index is often greater in plants that are shorter and produce fewer leaves, roots, and stems, concentrating their resources instead on fruits and seeds. Such plants may not survive as well in the wild, but would have been the favorites of the earliest agriculturalists. As a result of this process, the plant species would have been unconsciously bred for the invention of agriculture. When a brilliant man or woman then figured out how to raise the plants, agriculture was successful. For millennia, humans used both wild and domesticated grains as food.

Agriculture began independently in many parts of the world (table 10.1). The only major region in which agriculture did not evolve was Europe and possibly Australia. Agriculture began earliest (about ten thousand years ago) in the Middle East, especially in the Tigris-Euphrates floodplain of Mesopotamia. Agriculture began there first

TABLE 10.1. Centers and Times of Origin of Major Crop Plants

CENTER OF ORIGIN	CROP[a]	TIME OF ORIGIN (YEARS BEFORE PRESENT)
Africa	Sorghum	4,000
	Pearl millet	3,000
	African rice	2,000
Andes	Potato	7,000
	Quinoa	5,000
Central South America	Peanut	8,500
	Manioc	8,000
	Chili pepper	6,000
China	Millet	8,000
	Rice	8,000
Eastern North America	Squash	5,000
	Sunflower	5,000
	Chenopod	4,000
	Marsh elder	4,000
India	Mung bean	4,500
	Millet	4,000
Mexico	Squash	10,000
	Maize	9,000
	Bean	4,000
Middle East	Rye	13,000
	Fig	11,400
	Wheat	10,500
	Barley	10,000
New Guinea	Banana	7,000
	Taro	7,000
	Yam	7,000
Northern South America	Squash	10,000
	Arrowroot	9,000
	Lima bean	6,500
	Cotton	6,000
	Yam	6,000
	Sweet potato	4,500

SOURCE: Michael Balter, "Seeking Agriculture's Ancient Roots," *Science* 316 (2007): 1830–1835.

[a]Millet, squash, and yam are different species or varieties domesticated independently in different centers of origin.

apparently because there were many wild species of plants that were almost suited for agriculture in their wild state. As ecologist Jared Diamond has explained, of the fifty-six species of wild grains that have large seeds, thirty-two grow in the Middle East. These wild grains needed little evolutionary transformation to become crops. The transformation from gathering to agriculture would have been a relatively quick and easy process in that part of the world.[3]

This same process occurred in all the other places where agriculture originated, but it took longer. There were fewer suitable wild food plants in Mexico, and even fewer in the Andes. The development of agriculture was delayed in Mexico because a greater evolutionary transformation was needed to transform wild teosinte into domesticated maize.[4] It took still longer for poisonous wild potatoes to evolve into edible ones.

Once agriculture evolved, societies that depended on it could not easily revert to hunting and gathering (chapter 1). Agriculture allowed greater food production and greater population growth. These larger populations could no longer find enough food in the wild to support themselves. Because of the greater food production, it was no longer necessary for nearly all tribal members to participate in food procurement. Farmers raised enough food for everybody, which allowed other people to be soldiers and priests. A hunting, gathering tribe was ill-equipped to fight an agricultural tribe with a dedicated army. A world trapped in agriculture was trapped into war. Agriculture allowed the rise of religious and governmental hierarchies as well as of armies. At the bottom of the hierarchy were the slaves, who raised food for everybody.

With the evolution of agriculture, productive farmland became valuable. People settled into cities, because they had to stay in one place at least long enough for one harvest. Agriculture promoted the rise of civilization. Civilization arose earlier in Mesopotamia than in other places because agriculture began earlier there. Because they needed to defend particular tracts of territory, the armies now had a lot more to fight about. The cultural groups that developed agriculture first were the first to be civilized, and to have advanced technology, which allowed them to conquer the cultural groups in which this process had not progressed as far.

Agriculture did not improve human health. Because people lived in crowded, walled cities, trapped with one another, with their own wastes, and with rats, diseases were common. Agriculture actually decreased the quality of human nutrition by making people rely on a few crop plants rather than a diversity of wild foods. In particular, the human body evolved under conditions in which ascorbic acid (vitamin C) was readily available from wild fruits. When entire populations became dependent on crops with little vitamin C, scurvy became a way of life. Therefore, although agriculture caused the human population to *increase*, it also caused the average life span to *decrease*, compared with the already brutally short lives of hunter-gatherers. The lives of tribal hunter-gatherers are far from idyllic. But the emergence of agriculture did not create a paradise either. Despite the many benefits of agriculture, it is not an entirely pretty picture. But it is our picture—the origin of agriculture, and at the same time of the origin of civilization and history.

Agricultural Plants That Have Had a Major Impact on Human History

Although there are many kinds of agricultural plants, a few of them have had a particularly important impact on human history and the world economy. Four examples are sugarcane, grain crops, potatoes, and legumes.[5]

Sugarcane

Sugar from sugarcane (*Saccharum officinarum*), a tropical plant, was highly prized by late-medieval Europeans. Sugar and molasses were processed from the sap of sugarcane, and molasses could be made into rum. Sugar is not actually a food; it provides calories, but very little useful nutrition. Once Europeans got a taste for sugar (and rum), they couldn't get enough.

Because sugarcane will grow only in warm climates, sugar had to be imported into Europe. One of the motivations for early European explorations, along with gold and spices, was to find new land for growing sugarcane. Early Spanish and Portuguese explorers discovered the

Canary Islands and exterminated the native Guanche people; they then converted these islands, along with Madeira and the Azores, into sugarcane plantations. They cut down the forests to create cane fields and for the wood to boil the sap. Columbus wasted no time: on his second voyage, he brought sugarcane to Hispaniola.

The Spaniards under Columbus enslaved the Native Americans of Hispaniola to work in the cane fields. Within fifty years, the island's natives were extinct. Not about to let their cane fields and syrup mills go untended, and not about to do the work themselves, the Spaniards imported black slaves from West Africa. Many of the slaves perished, as had the natives, in the inhuman conditions of the cane fields and molasses refineries; but there were always more captives imported to replace them. European and American ships brought slaves from Africa to the Caribbean (the infamous "Middle Passage"), and the molasses from the Caribbean to England, where rum was distilled; the ships then returned to Africa to complete the last leg of the "Triangle Trade." Because so many slaves died raising sugarcane and processing molasses, slave importation into the Caribbean continued into the nineteenth century, even after it had stopped in the United States. Thus the sordid history of colonial slavery in the New World began with, and the worst slave conditions were found in, the fields that fed the sugar addiction of Europeans and, later, white Americans.[6]

Europeans continued to crave sugar even when import supplies were interrupted. The sugar beet (*Beta vulgaris*) was bred in Europe in the eighteenth century. Unlike sugarcane, beets could grow in Europe, so the product processed from them was not endangered by piracy or war on the high seas. Sugar makes people malnourished, because it displaces more nutritious foods in their diets, and its cultivation takes up some of the best tropical agricultural land. This is as true today, when peasants work in the cane fields of the Philippines and Brazil, as it was when slaves worked in the cane fields of the Caribbean. Despite this, sugar continues to be a leading agricultural product: sugarcane is the world's number one agricultural crop, in terms of gross production, surpassing the staple food crops wheat and rice. Sugar beets are also a major world crop.

Grain Crops

Sugarcane is a member of the grass family, Poaceae. Other members of this family are the most important foods in the world: wheat, rice, and maize. Wild wheat plants (genus *Triticum*) can still be found across the southern Mediterranean regions, such as Iraq and Turkey, where wheat was domesticated thousands of years ago. The gluten protein contained in wheat flour makes bread dough sticky and therefore able to hold bubbles of carbon dioxide produced by yeast. Called the "staff of life," wheat (*Triticum aestivum*) is one of the three major staple food crops in the world.

Rice is the major food grain for almost two billion people. More people depend on rice as a staple food than on any other crop. The species of rice most commonly grown, *Oryza sativa*, was domesticated in Asia thousands of years ago. A different species of rice was independently domesticated in Africa, also many centuries ago. North American wild rice (which has not been domesticated) is yet a different species. Many varieties of *Oryza sativa* are planted in flooded paddies. They are sown thickly, then later transplanted so that they have room to grow. Processing of brown rice into white rice removes much of the nutritious seed coats.

Maize (*Zea mays*) was domesticated by the Native Americans of Mexico long before Columbus arrived in America. Corn tassels consist of flowers that produce only pollen, while the ears consist of flowers that produce only seeds. The threads of silk are the structures on which pollen lands. Maize is a major American crop and a staple for survival in Africa. Maize is called "corn" by Americans, whereas in England "corn" refers to any grain.

Several other major crops are grasses as well: oats, rye, barley, and sorghum. Millions in Africa and Asia eat seeds of sorghum, a drought-tolerant grass, as their major food source. Rye and oats were weeds in the grain fields of prehistoric farms, before themselves being domesticated. Today, fields of wheat and barley account for nearly 40 percent of the agricultural land in the world.

Potatoes and Other Nightshade Crops

One of the most important agricultural crops in history has been the potato, which is a member of the poisonous nightshade family, Solanaceae. The variety of ways in which humans consume the potato—as chips, fried, mashed, or baked—masks an underlying genetic uniformity. In the homeland of the potato, the Andean highlands of Peru, native people may consume hundreds of varieties. Potatoes can grow at high elevations—they flourish up to 3,500 feet above sea level, higher than most other crops can survive—and they grow well in the poor, rocky soils that are frequently found in these places. Potatoes are propagated by growing new plants from tiny young *tubers* or underground stems, the edible portion of the potato plant.

The Spaniards, who conquered Peru, did not adopt the potato as a food, nor did most other European countries when they first encountered it. In Ireland, however, the story was different. The potato was introduced into Ireland and became the main food crop by 1625. By the middle of the nineteenth century, virtually all of Ireland depended on just one breed of potato. Many Irish soils were poor; in the outlying areas to which the English conquerors drove the Irish, no crop would grow well except the hardy potato. Poor Irish ate almost nothing but potatoes, but they could grow plenty of them. Partly as a result of the availability of so much food, the population of Ireland grew to be nine times greater than in medieval days.

In the eighteenth and early nineteenth centuries, several plagues struck the Irish potato crop, caused by various kinds of microbes. They caused locally severe famines. But in 1845 the entire island was struck by the Potato Famine. The microbe *Phytophthora infestans* (which causes *potato late blight*) was accidentally introduced from America. In South America, many varieties of potato had evolved resistance to this microbe. But in Ireland, the single variety of potato that was grown could not withstand the infestation. Although a healthy potato plant may not be immediately killed by this microbe, and it may not spread rapidly through a field of healthy potatoes, the disease can devastate fields of potatoes that have been made weak by bad weather, which is what Ireland experienced in 1845. The stage was set: because of weather and genetics, the potatoes were helpless before the newly introduced disease. Almost all the potatoes in the country melted into black putrefaction within a brief span of time.[7]

A million Irish people starved to death, or died in their weakened condition from disease, before 1846 was over. Even as the Irish were starving, good farmland, nearly all of it owned by Englishmen, continued to produce grain. British troops protected wagons of grain, destined for export, from the starving populace. Food was available in the markets, but at a price the victims of the famine could not afford. Several million Irish emigrated, many of them to the United States, where today there are more people of Irish descent than in Ireland. Ireland had a population of 8.2 million in 1851; by 1900, only 4.4 million remained.

The history of Europe consisted of an interminable series of wars. Some of these wars were prolonged and of low intensity, resulting more in the destruction of farmland than of troops. Starvation was a frequent consequence of war, as well as of bad weather. One major reason for this is that northern Europe, like southern Europe and the Middle East, depended largely on wheat. In order for a wheat crop to be successful, everything has to work out: suitable weather during the growing season, good weather at harvest, and not too many rats in the storage bins. Frequently, weather was bad, rats were abundant, and soldiers came along and burned the fields. Wheat was a fragile base for the food economy of the rural poor, especially in northern Europe where the weather was frequently unsuitable for this crop of Middle Eastern origin.

All of this changed when potatoes were introduced into northern Europe in the nineteenth century. The potato provided a more stable food base than wheat for the rural population. Potatoes grew better than wheat in the cooler conditions of northern Europe. And what damage could a hailstorm, or an army, do to a potato field? A field of potatoes, the aboveground parts of which are destroyed, will simply grow back from the tubers. Once the potato was widely adopted, rural populations of Germany and Russia grew. Some historians link the rise of the economies of Germany and Russia in the nineteenth century to the adoption of the potato as a staple crop by their populations.[8]

Several other members of the nightshade family have assumed worldwide importance. Native Americans in Mexico domesticated tomatoes and chili peppers. Can you imagine Italian cooking without tomatoes? That's what it was like before the 1500s. Can you imagine Thai or Szechuan food without chili peppers? That's what it was like until about the 1700s.[9]

Legumes

Leguminous plants (family Fabaceae) are important hosts for nitrogen-fixing bacteria, which contribute nitrogen to the soil when the plants grow and decompose (chapter 6), and allow the plants to produce large amounts of protein in their leaves and seeds. Soybeans (*Glycine max*), originally from China, are a major crop in the United States and Brazil. Though perfectly edible by humans in the form of soy sauce, tofu, or soy milk, most American soybeans are used for livestock feed and for oil extraction. The peanut (*Arachis hypogaea*), a native of North America, is a major crop in the United States, but even more so in western Africa, where peanut oil is produced for the European market. Peanut flowers are fertilized aboveground, then as the flower stem elongates the flower is forced underground where the fruits develop. Peas and lentils are also legumes.

Legume seeds are high in protein (soybeans can contain 38 percent protein) and oils (soybeans can contain up to 18 percent oil). Grains are insufficient by themselves as a source of protein because their proteins are deficient in lysine, one of the essential amino acids that human bodies cannot synthesize. The proteins of legumes are deficient in methionine, another essential amino acid. A mixture of grains and legumes, however, can provide an adequate balance of amino acids, in which lysine "complements" methionine. This is called *protein complementarity*. Many agricultural peoples have independently developed staple diets that contain both grains and legumes: maize and beans in

Central America, wheat and lentils in the Middle East, rice and soybeans in the Orient. "Man cannot live by bread alone" is true enough for modern wheat bread, but the bread traditionally eaten by peasants often contained bean flour as well as grain flour and was much more nutritious.[10]

Thus, the evolution of agriculture made civilization possible, and specific agricultural crops changed the course of human history. Agriculture has indeed changed the world. About 30 percent of the earth's surface is used for agricultural production. Farmland is more important than all but the largest natural habitats that were explored in chapter 8.

Agriculture Forever

Not only has agriculture been overwhelmingly important in human history, but it will remain just as significant in the future as it has been in the past. It is highly unlikely that we will ever raise food for billions of people in hydroponic greenhouses, away from the soil and the land— not because of technological limitations but because of the cost. Yet even in a greenhouse, it would be mostly the old, standard agricultural crops that would produce our food. There will never be a day when little green algae will feed everybody, not because we cannot do this, but because nobody wants to eat little green algae. We will, forever, rely on many of the crop plants that the earliest farmers left us. Nevertheless, farmland is being degraded at the same time that the world population continues to increase. We must therefore transform agriculture into a form that fits into the natural systems of the earth. Our current system of crop production, environmentally destructive and dependent on the expenditure of fossil fuels, does not do so.

According to the epigraph at the beginning of this chapter, the destruction of forests was one of the few things that gave Thomas Jefferson second thoughts about democracy. And yet, as the quote suggests, it seemed necessary to cut the trees down in order to create farmland. How could Jefferson preserve the old forests as well as the landscape of yeoman farmers, both of which he considered necessary to the future of the nation? This seeming contradiction is based on the assumption that agriculture requires the complete elimination of the natural landscape within the farming region. But centuries of experience and current agricultural research both suggest that the best farms are those that incorporate the natural world. Two systems that do this are natural systems agriculture (chapters 6, 7, 11), which uses mixtures of perennial crops, and agroforestry, described here.

Agroforestry is a system in which crops grow in conjunction with trees and shrubs. Trees and shrubs are not obstacles that need to be removed in order to plant a farm; rather, they can provide direct benefits to the farmland if we allow them to grow as borders, or fencerows, around the fields. The trees may be agriculturally important species that produce fruits, nuts, or other products, or they may be wild trees. Agroforestry makes sense from the forestry viewpoint as well. It takes decades for a tree plantation to be profitable, but when agroforestry is implemented, the crops provide immediate profit, and the trees pay off later.[11]

One benefit that fencerow trees offer to agriculture is that they can help to prevent erosion. The trees slow the wind near the ground, and as a result, it carries away less topsoil. The trees also retard the flow of water and the erosion that it causes. These advantages were explained in general terms in chapters 5 and 7; we now see that fencerow trees can provide such benefits directly to farms. In addition, fencerow trees supply shelter for birds and for predatory insects such as wasps. These predators serve their own interests by hunting crop pests, such as beetles and caterpillars, and thereby serve the interests of the farmer and human society as well. But they do not travel very far from their shelter. They cannot protect a square mile of treeless farmland from pests.

Farmers usually use pesticides to control pest outbreaks, but pesticides have numerous hazards: they contribute residues in food and water that are poisonous to people as well as to insects; insect populations evolve resistance to them; and they concentrate through the food chain. Natural predators pose none of these dangers. Besides, pesticides may in some cases kill the natural predators even more effectively than they do the pests, because the pests, with their large population sizes and short generation times, can evolve resistance to these chemicals more rapidly than the predators can.

The benefits of fencerow trees and shrubs have long been recognized. Before the invention of barbed wire, it was common in the American Midwest for farmers to plant thick hedges of thorny *Maclura pomifera*, known as Osage orange, hedge-apple, or bois d'arc. The thorns discouraged the movement of grazing animals. Bois d'arcs also did not grow tall enough to cast extensive shade on the fields. A native primarily of the Red River watershed of Oklahoma, Texas, Arkansas, and Louisiana, this tree was planted throughout the eastern United States. Even after barbed wire became common, fencerows remained common practice because they reduced erosion. It is not always necessary to intentionally plant fencerow trees and shrubs. Many of them, such as black cherries and blackberries, grow readily from seeds excreted by birds that briefly sit on the fences, whereas others, such as cottonwood, grow from windblown seeds that settle in the little spaces of calm air near the fences. Fencerows therefore represent a free service to agriculture.

Beginning in the 1970s, many fencerows were removed, under the encouragement of U.S. Secretary of Agriculture Earl Butz, who told farmers they should plant crops "fencerow to fencerow"—and he did not mean fencerow trees. A 10-foot swath of trees represents a considerable amount of farmland, and a farmer can grow more crops when this land is cultivated. This practice made little sense in a market characterized by overproduction and government subsidies that are designed to keep some land out of production.

Most agricultural research emphasizes large fields and the massive use of pesticides, partly because these practices are in the economic interests of large corporations. If a farmer saves money because fencerow trees reduce the need for pesticides, the farmer benefits but the corporations that produce the pesticides do not. A small but growing amount of agricultural research, however, is investigating the use of natural pest control. The birds and predatory insects control crop pests by feeding themselves, and trees control farm erosion simply by growing. Jefferson would have been pleased to see that modern science has revealed that the felling of all of the trees is not, in fact, necessary for the production of food, and that trees may hold the key to the success of agriculture in the future.

chapter eleven

WHY WE NEED PLANT DIVERSITY

Nature has introduced great variety into the landscape,
but man has displayed a passion for simplifying it.
—RACHEL CARSON

Representative James V. Hansen called it "a shot across the bow from a retiring chairman" and apparently considered it one of his best. Hansen, the Republican representative from Utah's First District and chairman of the House Committee on Natural Resources, filed a "landmark bill" in late 2002 that would exempt private property, military lands, and *all plant life* from the Endangered Species Act.[1] The Endangered Species Act has been, since its inception, the law that conservatives have loved to hate. Why, they ask, should "the rights of an endangered fly or a species of seaweed," in Hansen's words, interfere with property rights and the investments of industry and agribusiness? In fact, this law has caused the cancellation of only a small percentage of construction contracts. The examples most often cited are those with unfortunate names; for, it must be admitted, the furbish lousewort (a small wildflower) hardly sounds like something to stir public sympathy. Conservatives have been successful. The period between May 2006 and May 2007 was the first time since 1981 in which an entire year passed with no new species being listed as endangered.

There is immense public support for the protection of what some writers have called "charismatic megafauna": grizzly bears, wolves, bald eagles, bison, elk, pronghorns, and moose, just to mention the North American examples. No politician wants to be seen as the enemy of these popular animals. Indeed, Hansen mentioned the first three of the creatures in that list as the things that the Endangered Species Act was meant to protect in the first place. He apparently considered the world of humans and of endangered species to be a zero-sum game: that is, if they win, we lose; there is not enough room for them and for us, at least on private property and military bases. He said: "If we exempt private property, military lands, and all plants from the ESA, we would, in short order, have a more prosperous and secure nation."

As outrageous as his statement was in the eyes of conservation organizations all over the country, Hansen actually raised a good point—one that I brought up in the introduction to this book. Plants are important, but why should we care about preserving *every species* of plant? The answer to this question is very clear, but not obvious. The preceding chapters were about the role of plants, in general, in keeping Earth alive and, in the human economy, about the importance of the major agricultural plant species. This chapter explains three reasons we need to protect not just plants in general but the entire diversity of plant species: as sources of medicine, as the foundation of agriculture, and as an essential part of the natural world.

Wild Plant Species as Sources of Medicine

Wild plants have been and will continue to be important sources of medicines. The leaves of wild plants produce toxins that deter fungi and herbivorous animals from damaging them. The world is not a big salad bowl, a fact we sometimes forget, as we are accustomed to encountering garden plants that have been bred to be delicious. Many of these toxins, when used in small quantities, have medicinal effects. Most medicines began as natural plant extracts. For example, aspirin, though today synthesized artificially as acetylsalicylic acid, was once extracted from willow bark and received its name because it was first discovered in a wild spiraea bush. The plants of the tropical forest are an especially rich source of medicines, because their leaves must be very toxic to protect themselves from the year-round attack by insects and other animals. Wild plant communities, particularly tropical forests, are rich laboratories for the production of medicinal compounds, as millions of plants in thousands of species are busily producing new genetic variations of chemicals. Pharmaceutical research corporations often base their research on plant extracts. Because each kind of plant has a unique combination of protective chemicals, we need to save all of them. The extinction of any wild plant species, such as the seaside alder described in the introduction, might mean the loss of a yet-undiscovered medicinal compound.

There is no way to predict, beforehand, which wild plants might be "worth saving" for medicine or for some other useful product. As I and other scientists have found in our research into the pharmaceutical uses of wild plants, you cannot guess which plant extracts may be effective against which medical condition. Who could ever have guessed that certain forms of leukemia could be treated with vincristine and vinblastine found in the leaves of *Catharanthus roseus*, a little pink wildflower that lives in the vanishing rain forests of Madagascar? Or that ovarian cancer could be treated by taxol from the bark of a little bush, *Taxus brevifolia*, from the north Pacific forests? One promising way is to listen to the lore of tribal peoples who have had millennia of experience with medicinal plants. They can, at least, point the way toward finding plant species that have potent properties worthy of investigation. *Ethnobotanists* are scientists who visit tribal peoples and ask them about their native traditions concerning medicinal plants.[2]

Meanwhile, as tribal peoples are incorporated into modern civilization, they begin to depend more and more on the products of modern agriculture and industry. Tribal children in rural areas can identify hundreds of wild plants and know what they are useful for; tribal children in urban areas, like most children in highly industrial countries, know very few plants. Tribal knowledge is vanishing much more rapidly than are the tribes themselves. There are many thousands of members of the Cherokee tribe in the United States (I am one), but almost none of us know any of the botanical lore that was a normal part of life in our tribe a couple of centuries ago. I learned all of my botany in college, not from my ancestors.

Wild Plant Species for Food and Agriculture

When humans lived as hunters and gatherers, which we did for more than 90 percent of our history, we ate a great diversity of foods, including wild seeds and fruits. The physiology of our bodies evolved a dependence on this diversity of food sources. For example, some mammals have can synthesize ascorbic acid, a molecule that is necessary in the formation of certain tissues in their bodies. But the ancestors of modern humans lost this ability. This was not a problem, however, as ascorbic acid (also known as vitamin C) was abundant in the wild fruits that primitive humans ate. Not just plants, but a diversity of plants, has always been important in our diet.

Vitamin C deficiency became a problem after the advent of civilization, as many of the poorest humans had to survive on a bland diet of grains. They suffered the symptoms of scurvy as a result. Fortunately, agriculture supplies us not with just the basic staples, such as wheat, but with a large number of other foods, such as apples and oranges, which supply nutrients like vitamin C that the staples cannot provide. Agriculture can meet the full range of our nutritional needs, not just our need for calories. The database of the United Nations Food and Agriculture Organization (FAO) lists 152 different major crops produced in the world.[3] Even though only a dozen or so crops constitute the lion's share of agricultural production, our health and our economy would suffer substantially without the others. Our health continues to depend on plant diversity—the diversity of crop species.

The new science of evolutionary medicine has revealed that human health may be best served by a diet that most closely resembles what our hunter-gatherer ancestors ate. In particular, disorders such as diabetes appear to be stimulated by a reliance on modern, processed foods. If everyone were to go back to a traditional, diverse diet, it would seem that 152 kinds of crops would provide sufficient variety, without the need to gather wild foods or domesticate new kinds of crops. However, as botanist Gary Paul Nabhan has explained, there is no single primitive diet to which all humans have adapted. He notes that certain genetic lineages of people, such as Native Americans, Native Hawaiians, and

Native Australians, suffer the most from diseases such as diabetes when they depart from traditional cuisines and adopt modern culinary uniformity. He proposed that, during the course of the past few thousand years, cultural groups have biologically adapted to consuming local foods. The wild succulent plants consumed by desert peoples, he points out, not only have less of the quickly metabolized starches that contribute to diabetes, but also contain more of the molecules that help them to resist diabetes.[4] If a close coevolution has occurred between local people and their local crops and wild plants, then 152 kinds of plant-based foods are not enough for the optimal health of all humans.

But the story of diversity does not stop with plant species diversity. We need not only a great diversity of crop plant *species*, but we also need a great genetic diversity of *breeds* or *varieties* within each crop species. There are scores or hundreds of breeds of each species of crop plant. Many of the characteristics of each species are encoded in *genes* within the DNA of each cell. Genes are passed down from one generation to the next. Even though all of the members of one species may have the same genes, they have *different versions* of those genes. These different versions are known as *alleles*. Moreover, the DNA of each individual has *control sequences* that determine which genes are put to use, and to what extent. This is as true for crop species as it is for humans. Crop breeds differ from one another because they have different alleles and because they have different control sequences. That is why there are so many different breeds of corn, or tomato, or apple.

Plant breeders, as the term might suggest, spend their entire careers producing yet more breeds of crop plants. Traditionally, they have done so by cross-breeding existing varieties of crops to produce new combinations of characteristics. More recently, the techniques of biotechnology have allowed plant breeders to move the DNA from one breed of plant to another. Because the DNA of all species is chemically the same, biotechnologists can move genes from one species of plant to another, or even insert a viral, bacterial, or animal gene into plant DNA. Thousands of acres are now devoted to crops that contain viral or bacterial genes.

The importance of maintaining a diversity of crop breeds is not immediately apparent to most people. Although it is nice to have different breeds of apples from which to choose, just how important is it? After all, these breeds do not differ greatly in their nutritional value to humans. The importance of having a great diversity of crop breeds is not so much for everyone's pleasure in eating them as it is for farmers' success in growing them. We need to preserve a great diversity of crop breeds because they each have adaptations to different climates and to different biological environments. Some breeds of wheat tolerate cold winters or droughts better than others, and some breeds of corn can complete their life cycle in a shorter period of time than others, allowing them to grow further north where the growing season is shorter. These are examples of adaptation to different climates. Adaptations to different biological environments, which consist of the other plants, the viruses, the fungi, and the insects that the crop encounters, are even more important. Some crop breeds are better than others at growing in the crowded conditions of modern agricultural fields. Some can resist viruses, fungi, or insect pests that others cannot. Plant breeders have to keep producing new breeds because, once an established breed has been grown for a long time, viruses or bacteria or fungi will evolve the ability to infect it. It is an evolutionary race between the natural evolution of the pests and the artificial evolution carried out by plant breeders. Breeders keep reshuffling the alleles of different crop varieties to produce new adaptations to changing climate and to new pests.

Neither traditional plant breeders nor modern biotechnologists create new alleles—they just move them around. They have to obtain the alleles from existing populations of organisms. And if those organisms become extinct, the alleles that they contain vanish and are no longer available for the production of new breeds of crops.

Three Threats to Plant Diversity in Agriculture

There are three kinds of extinction that threaten the future of agriculture, as well as the development of new commercial products. The first is the extinction of *traditional varieties* of crops, often called "heirloom" varieties. All around the world, subsistence farmers and gardeners have developed numerous breeds of crop that are well-suited to their local conditions and the preferences of their cultures. They pass the seeds of these varieties from one generation to another. This is why the world contains hundreds of varieties of major crops such as wheat and minor ones such as chili peppers and eggplants. But often these heirloom varieties have inferior yield in comparison to the varieties sold by the large seed companies of Europe and North America. Subsistence farmers may, if they can afford to do so, choose to eat their old seeds rather than save them back to plant the next year, and buy the varieties that they perceive as superior. The old varieties, and the benefit that they may offer to the future of plant breeding, may be lost as a result. By eating rather than planting the last seeds of the old varieties, they are literally eating the future. As Barbara Kingsolver has written, we "rely on the gigantic insurance policy provided by the genetic variability in the land races [of crops], which continue to be hand-sown and harvested, year in and year out, by farmers in those mostly poor places from which our crops arose."[5]

Plant breeders do not want these traditional varieties of crops to become extinct, and some governments are trying to rescue them before they vanish. The U.S. Department of Agriculture (USDA), for example, has cold-temperature storage facilities for major crops, such as the one for potato varieties in Sturgeon Bay, Wisconsin. But such facilities are expensive, and they are used only to save the varieties of the crops that earn the most money for large-scale agriculture. Because cold-temperature storage in warm countries is expensive, five northern countries (led by Norway) established a seed vault in the arctic Svalbard Islands in 2006.[6] The facility will eventually have three million seeds and store more varieties of more crops than any previous project. But these measures by themselves are not enough. Seeds will die after being stored for a few decades, so they must be planted and grown to produce new seeds, which are then stored again.

Because the USDA focuses primarily on the major crops, it is largely up to private organizations, such as Seed Savers Exchange in Iowa and Southern Exposure Seed Exchange in Virginia, to rescue the varieties of minor crops like bell peppers and okra. These organizations cannot afford low-temperature facilities. Instead, they send old seeds to private gardeners, who raise them and send back new seeds.[7]

By preserving the rare varieties of major crop plants, plant breeders are ensuring the future survival of these crops. Individuals and private organizations do the same for the minor crops. Often, individuals save varieties not just for their usefulness but for pleasure. In addition to the regular kind of basil with which most aficionados of Italian cuisine are familiar, there are many other varieties, which have flavors similar to everything from cinnamon to lemon. There is also a garden variety of "chocolate mint" whose raw leaves live up to their name. Diversity within major and minor crop species is important for both practical reasons and for sheer enjoyment.

The second kind of genetic extinction is of the *wild relatives* of crop plants. Wild breeds or species of plants that can cross-breed with crop plants can provide even more genetic variety than the heirloom crop varieties. For example, there are at least 212 species of cultivated plants from the tropics alone that have wild populations closely related to them. Chocolate, the health benefits of which are now being revealed, is *Theobroma cacao*. The genus *Theobroma* has twenty-one wild species that can be useful in producing new breeds of cocoa.[8] There is at least one famous example of wild species rescuing a cultivated one from danger. When phylloxera bug infestations threatened European grapevines (*Vitis vinifera*) and the wine industry in the late nineteenth century, botanist T. V. Munson found that wild grapes (*Vitis rotundifolia*) from Oklahoma and Texas could resist this insect. He shipped thousands of wild grape rootstocks to France, and French grapevines were grafted to these roots. This action is reputed to have saved the wine industry.[9]

But like the traditional garden varieties, the wild relatives of crops are vanishing, mostly because the natural habitats in which they live are being destroyed. The wild ancestors of wheat, for instance, persist in the hills of the Middle East, where war and land degradation endanger them. Some species of teosinte, the wild ancestor of maize, barely survive in the foothills of Mexico. In some cases, traditional garden varieties have avoided extinction by escaping into the wild. For example, wild coffee trees live in the forests of Ethiopia adjacent to the fields of cultivated coffee, but these forests, like most others, are shrinking under human influence. For a few crops, such as mangoes and papayas, the wild relatives thrive in disturbed areas and are therefore in no danger of extinction; such examples, however, are in the minority. And for even these species, at least some of the wild relatives are threatened with oblivion.[10]

The third kind of extinction that jeopardizes the genetic basis of agriculture is the extinction of *possible new crops*: wild plants that might, in the future, be domesticated. There are many wild species of plants that offer the promise of future development as crops. Given that we already have so many kinds of crops (even more than the 152 major categories listed by the FAO), why do we need to develop new ones? This is a good question, especially because the development of a new crop is a protracted and difficult process. But there are definite economic and ecological benefits to the development of new crops from wild species. Wild plants are not only important for individual survival in an emergency—the U.S. Army Survival Manual describes seventy-four of them[11]—but for the future of agriculture.

New crops may offer significant market possibilities if their taste or usefulness is unique. For example, most persimmons sold in the United States are Japanese persimmons (*Diospyros kaki*) but the wild American persimmon (*Diospyros virginiana*) provides a fruit whose wonderful flavor differs not only from all others in the supermarket but from the Japanese species as well. Likewise, there is no other fruit with a flavor quite like that of the mulberry, *Morus alba*. A significant amount of work would be necessary to prepare these two wild species for orchard cultivation. The American persimmon fruit is small and full of large seeds, and mulberries typically lose their flavor if frozen. Wild fruits usually ripen over a prolonged period that would make simultaneous harvest impractical. Still, someday we may find that there is a market niche for them. Other wild plants, such as dandelions and violets, and even (after cooking) nettles, are eminently edible and nutritious.[12]

Perhaps more important, new crops may offer significant ecological advantages. Some might, for example, be able to grow exceptionally well in degraded or dry soils otherwise unsuitable for agriculture.

The spread of soil degradation, through erosion and salinization (chapters 5 and 7), leaves us three possibilities: (1) slow down soil degradation, (2) develop new crops that can live in degraded soil, or (3) give up. It is nice to have two, rather than one, alternative to giving up. There are three ecological circumstances for which new crops might offer a partial solution.

New crops for saline soils. Relatively few crop plants can grow in salty soil (beets and barley are among the few exceptions), and even these cannot tolerate very much salt. However, some wild plants can tolerate high levels of salt. Wildwheat (*Distichlis palmeri*), a wild grass native to the southwestern United States, produces an edible grain; and the oil-rich seeds of the pickleweed *Salicornia* can be added to animal feed.

New crops for dry soils. Some wild plants that thrive under desert conditions might prove useful for agriculture in the future. The jojoba (*Simmondsia chinensis*), a bush from the North American southwest desert, produces a liquid wax that can be useful in industrial processes because it is stable under high temperature and pressure. Many acres of jojoba have been raised in irrigated lands in the desert southwest. Although the bush does not need irrigation, water supplementation boosts its growth and yields. Unfortunately, this bush became a victim of its own success. Irrigation raised the land values so much that jojoba plantations, which yield only a moderate profit, have been converted to lucrative subdivisions. Other arid land plants, however, grow well without irrigation. Prickly pear cactus (*Opuntia ficus-indica*) produces edible fruits (*tuna* in Spanish) and edible pads or stems (*nopalitos* in Spanish). Buffalo gourds (*Cucurbita foetidissima*) grow well on roadsides of the deserts and high plains. As its Latin name suggests, it is inedible; however, its massive root can yield starch for ethanol production. Its seeds can produce more oil than some cultivated oil crops, with no irrigation.[13]

New crops for polluted areas. Wild cattails can grow in water contaminated by sewage. The bacteria that grow around their stems and roots break down the sewage, and the cattail plants absorb the released nutrients. Arcata, California, uses cattail marshes for its municipal sewage treatment.[14] The underground stems of the plants produce a great deal of starch, which can be harvested for the production of fuel ethanol. Some agricultural researchers are investigating the possibility of promoting hazelnut bushes, which grow rapidly and produce a highly marketable product, to control runoff from (as well as concealment for) hog confinement facilities.

Plant breeders can also produce new crops that cause less environmental degradation in the first place. Perhaps the best example of this is natural systems agriculture (NSA), which is the focus of research at the Land Institute in Salina, Kansas (chapter 6). NSA uses perennial rather than annual crops and therefore requires less energy and less fertilizer, as well as minimizing soil erosion. Very few of today's main agricultural crops, among them potatoes and alfalfa, are perennials. In order to promote an agriculture based on perennials, Land Institute scientists have to develop new crops. And to do so, they draw from the resources of wild plant species.[15] For example, wild wheatgrass (*Thinopyrum intermedium*) is a perennial relative of wheat that could be used as a source of grains. Maximilian sunflower (*Helianthus maximiliani*) and rosinweed (*Silphium integrifolium*) are perennials that could be used as a source of oilseeds. Bundleflower (*Desmanthus illinoensis*) is a perennial legume that produces seeds similar to those of soybean, only smaller.

The main work of plant breeders at the Land Institute is to cross wild perennial plants with annual crop species. This is a quicker, and more promising, route to the development of perennial replacements for annual crops than is the direct breeding of wild perennials. The researchers cross perennial wheatgrass with annual wheat, Maximilian sunflower with annual sunflower, and perennial Johnson grass (*Sorghum halepense*) with annual sorghum (*S. bicolor*). They then cross these hybrids with the annual parents (fig. 11.1), thus producing plants that more closely resemble the annual crops, except that they may be perennials. They plant thousands of cross-bred specimens, and then see which ones grow for more than a year. From among these survivors, they choose the plants with the best characteristics. The best characteristics are usually the same ones that the earliest farmers selected: breeds that have large seeds, do not shatter the mature seeds, have no seed dormancy, have a high harvest index, and do not collapse under the weight of their own seeds (chapter 10).

FIGURE 11.1. A researcher at the Land Institute cross-breeds sunflowers as part of a project to develop perennial sunflowers as an oil crop. Photograph by the author.

As we stood in a field of sunflower hybrids, Land Institute researcher Dave Van Tassel showed me the tremendous diversity of plant characteristics: some had single tall stalks, some had multiple shorter ones; some had a few large heads of seeds, some had more and smaller heads. The researchers save the crosses that have the highest yield and use them for further breeding. It may take many plant generations of breeding before new perennial crops can be developed; on the other hand, just the right cross may produce, in a single generation, a hybrid that has the right characteristics. If successful, perennial crops will allow farmers to produce grains, oilseeds, and legumes that have characteristics desirable to humans, that minimize soil erosion, and that promote fertility of the soil and its ability to hold water.

Saving old crops can be as useful as finding new ones. Native Americans of Mexico and South America grew large amounts of amaranth and quinoa, the seeds of which can be used as grains even though they are not grasses. The conquistadores tried to stamp out the cultivation of these crops, ostensibly because of their association with paganism. Fortunately, they were not entirely successful. Although they are not major crops, amaranth and quinoa are still grown and eaten in the places where they once flourished. You can even buy them in some American supermarkets. Many of the traditional grains of Africa, such as an African species of rice, finger and pearl millets, fonio, sorghum, and tef, are still grown in rural areas.[16]

Plant Diversity down on the Farm

The importance of plant diversity is perhaps most obvious not in what we grow on our farms but the way we grow it. Modern agriculture is overwhelmingly based on *monoculture*, in which farmers plant large expanses of just a single breed of a single species of crop plant. Monoculture invites all kinds of problems. For example, pests and diseases can spread rapidly through a monoculture of crops, because their preferred food plant is always just a few inches away. If a whole field consists of one species, or even one variety, of crop, all of them are competing for the same nutrients in the same soil level, the same sunlight at the same height above the ground. Despite these problems, farmers in industrialized nations usually grow their crops in monocultures, because a huge, genetically uniform field is easier to manage by means of machinery and chemicals. If a field contains nothing but corn and all of the ears of corn are exactly the same height above the ground, a mechanical harvester can get all of the corn by just driving back and forth across the field.

One of the worst examples is the modern banana. Thousands of square miles of land in Latin America are a monoculture of just one kind of banana: the Cavendish. Because domesticated bananas do not produce seeds, they must be propagated by cuttings, all of which are genetically identical. A genetically uniform monoculture is an invitation for the spread of disease. This is precisely what has happened: Panama disease, caused by a fungus, *Fusarium oxysporum*, has devastated banana plantations in southeast Asia, and it is only a matter of time before it spreads to

Latin America. Fruit companies first marketed the Cavendish banana because the previously popular variety, the Gros Michel, had been devastated by a different strain of Panama disease.[17]

Planting more than one kind of crop in the same field, a practice called *polyculture*, can greatly reduce these problems. Because each of the crop species is at a lower density in a polyculture than it would be in a monoculture, pests spread much more slowly through fields that consist of more than one crop species. One example that I saw in California was rows of sunflowers between fields of tomatoes. The sunflowers can impede the movement of sphinx moths through tomato fields, causing them to lay fewer eggs, which hatch into voracious tomato hornworms. Polyculture is an essential component of natural systems agriculture. By using perennial plants in polyculture, NSA is following the lead of geneticist Wes Jackson, founder of the Land Institute, who takes the tallgrass prairie as his inspiration and model. A mixture of perennial species has made the prairie a successful natural habitat; the same system should work for agricultural lands. His vision is to make agriculture incorporate natural processes that have been perfected by millions of years of evolution.[18]

If the crops in the mixture are of different sizes, they might be able to use the sunlight and the different soil depths more efficiently than could a monoculture. The gardens of tribal peoples almost always contained mixtures of species growing side by side (fig. 11.2). Native American gardens were centered on the "three sisters"—maize, beans, and squash. The beans and squash grew near the ground and also used the tall corn as a trellis. Other examples include growing rows of papaya trees in pineapple fields in the Philippines[19] and growing orange trees underneath date palms in southeastern California. The papaya trees do not shade the pineapples, nor do the date palms shade the oranges, enough to impede their growth. In fact, some crops such as coffee bushes grow better in moderate shade.

Another way in which polyculture can enhance the efficiency of agricultural production is that a farmer can benefit from raising more than one product. Although this may not particularly matter in the market economies of industrialized countries, it can be very important in rural

FIGURE 11.2. In the reconstructed pre-1830 Cherokee Nation capital of New Echota, in Georgia, a typical garden has the "three sisters" mixture of corn, squash (in this case, watermelons), and beans. Gourds provide nests for birds, which eat insect pests. Photograph by the author.

areas of the developing world. For example, farmers who plant the fast-growing tropical tree leucaena (*Leucaena leucocephala*) along with their primary crop can harvest the crop as well as the leucaena. Because the trees are leguminous, they have nitrogen-fixing bacteria in their roots, which allows their pods and leaves to be nutritious food for livestock, and even means that they can enrich the soil with nitrogen fertilizer (chapter 10). In addition, the trees can help prevent soil erosion.

There is no reason that polyculture cannot be integrated into modern mechanized agriculture. Instead of having a square mile of corn, and another square mile of soybeans, a farmer can plant alternating swaths of corn and soybeans that are still wide enough to allow farm equipment to maneuver. Such fields used to be a rare sight but are increasingly common in the midwestern United States.

Some people have the idea that all the wild plants that could be domesticated have already been brought under cultivation. Land Institute researcher Dave Van Tassel, among many others, rejects this notion. Once agriculture evolved, he told me, it was easier for farmers who moved to new locations to bring crops with them than to develop crops from wild plants in the new location. The success of the early crop plants therefore preempted the development of equally promising, or even superior, new crop plants. The great diversity of wild plant species therefore still represents a cornucopia of possible new resources for agriculture.

Plant Diversity in Nature

Finally, it is essential to save as many plant species as possible because the natural world depends on it. Each species of plant plays a unique role in the web of ecological interactions, a role that no other plant species can fill. For example, some animals eat only one kind of plant. Most people are familiar with the fact that pandas eat only various kinds of bamboo and koalas eat only a few varieties of eucalyptus. Extinction of one kind of plant can cause the extinction of animals that rely on it. A forest or a prairie, with high species diversity, can resist the spread of diseases better than tree plantations, for the same reason that polycultures can resist disease better than monocultures.

Nature is not fragile, in the sense that the disappearance of one species causes a whole cascade of other extinctions. When a fungus killed all the wild adult chestnut trees (*Castanea americana*) in the early twentieth century, various species of oaks filled in the gaps. But the web of ecological relationships is complex enough that we cannot know the consequences of the extinction of any species. Without doubt, however, the extinction of very many species will cause the collapse of the community of organisms—and that collapse would probably not be gradual. It is inevitable that we will lose some plant species. We must try to save *all* of them, so that we may succeed in saving *most*.

Plant diversity is as much of a blessing from the plant world as is the production of oxygen and food calories, the regulation of carbon dioxide, and the protection of water and soil. The world of plants is a vast cornucopia of genes that provides for every current need and may potentially provide for many future needs as well. Also important is the intelligence of observant traditional farmers and gardeners who, all over the world for thousands of years, have chosen wild plants and developed them into an astonishing variety of breeds within each of a vast number of crop species. Not only the loss of plant species and varieties, but the loss of cultural knowledge about these plants and how to grow them, is a major hemorrhage from which modern society suffers.

chapter twelve

WHAT CAN WE DO?

It is difficult to get a man to
understand something when
his salary depends upon his
not understanding it.
—UPTON SINCLAIR

Most of the ecological problems facing the world are interconnected with one another. This is both bad and good news. The
bad news is that we cannot make much of a difference in the world by
solving just one of the ecological problems, such as the greenhouse
effect, or soil erosion, or the depletion of water supplies; we have to
solve all of them. Compounding these difficulties is the fact that ecological problems are interconnected with economic problems, in particular the poverty of billions of people in the developing world. But the
good news is that by solving one problem, we will simultaneously help
solve the others. Saving the wild forests and preserving trees in the areas
in which we build houses and offices not only helps to control global
warming but also saves energy, slows down erosion, and enhances water
supplies. Reducing our use of paper and metals will not only slow the
greenhouse effect but will also reduce the impact of lumbering and
mining, which typically destroy natural habitats. Planting trees provides
direct economic benefits to poor people throughout the world. Wangari
Maathai, the Nobel Peace Prize winner, believes that dealing with ecological issues is not an alternative to solving the problems of poverty;
instead, it is an essential part of solving the problems of poverty. She
says that watching a poor person cut down the last tree for firewood to

cook her last meal is a situation we need to avoid by planting trees now.[1] She has demonstrated that getting millions of people to plant trees results in not only environmental, but social and economic, benefits that far exceed the actual contribution of the trees.

As explained in preceding chapters, planting trees is not going to save the world from the stresses to which modern civilization is subjecting it, whether they be the greenhouse effect, flooding, or soil erosion. Planting as many trees as possible and saving the forests we now have will help a great deal, but it will not be enough. What else do we need to do? There is no single list of things that we should all do, but here are some examples.

One obvious choice we could make is to *drive fuel-efficient vehicles*. For the past seventeen years, I have driven vehicles that get more than 40 miles per gallon: two successive Geo Metros and now a Toyota Echo. I have driven more than 300,000 miles in these cars, and in doing so I have used about 7,500 fewer gallons of gasoline than I would have burnt by driving a typical 20-mile-per-gallon vehicle (incidentally, saving me more than $22,500 in fuel costs at 2008 prices). I have also put only half as much carbon dioxide into the air as I otherwise would have done. I have found that driving fuel-efficient cars is not a particularly heroic or sacrificial thing to do. Hybrid vehicles get even better mileage, but if Americans would increase the average fuel efficiency by just a few miles per gallon, there would be a vast decrease in American carbon emissions even without a shift to hybrid vehicles. Cars made in the United States have the world's worst fuel efficiency, about 23 miles per gallon. Fuel efficiency of European cars has increased from about 37 miles per gallon in 2002 to almost 42 miles per gallon in 2007. Fuel efficiency of Japanese cars has remained steady at near 47 miles per gallon, about twice the American average.[2] The United States has 30 percent of the world's vehicles and produces 45 percent of the world's vehicle carbon emissions; what we do with our cars will make a tremendous difference in the world. The U.S. government goes so far as to say that increasing fuel economy will enhance national security by decreasing dependence on petroleum imports.[3]

Another obvious thing to do is to *use energy more efficiently* at home and at work.[4] Perhaps the hottest little topic of environmental discussion in 2007 has been the widespread adoption of compact fluorescent (CF) light bulbs. CF bulbs are more expensive than incandescent bulbs, but they last much longer and use only half as much electricity, with the result that they end up saving money for the consumer over the course of several years. An important consequence of saving energy is that utilities will not need to build as many new power plants as the population grows. California, Australia, and many cities have decided to make the transition in their government buildings and to offer incentives for consumers to do the same.

Good insulation, and the use of solar heating, can also reduce the energy that a home must use from commercial sources. Trees and other plants, whether around a building or on "green roofs," have proven successful at saving energy (chapter 4). Buildings all around the country are already using ice for air conditioning. Large freezers make ice at night when electricity demands are lower, then use the ice to partially cool the air during the daytime.[5] The wood-burning stoves of billions of poor people around the world can be modified, with appropriate help from governments and aid organizations, to burn wood more efficiently, reducing carbon emissions and saving poor women countless hours of looking for and carrying firewood.

Recycling often saves energy because processing new materials requires more energy than recycling old materials. This is especially the case with aluminum; recycling aluminum uses ten times less energy than does the processing of raw bauxite ore. Using recycled paper also saves energy, but not as much. The use of recycled paper at least does not require the felling of more trees. We published this book on recycled paper, even though recycled paper costs much more than paper made from new wood. This is attributable largely to subsidies that the federal government gives to lumber companies: the U.S. Forest Service allows lumber companies cheap access to trees and even builds roads for them. Paper recyclers receive no analogous subsidies. I do what I can to reduce paper consumption. Whenever possible, I print on the backs of previously-used computer paper. You may have heard the phrase "Reduce, reuse, recycle," without realizing that the order of the verbs is important: reducing the use of materials saves more energy than reusing them, which saves more energy than recycling them, which saves more energy than starting from scratch.

Many *alternative energy sources* are also ready to use. Sunlight, wind, and geothermal energy release no carbon dioxide into the air; it is no surprise that there is a rapidly growing market for them. They are available, now, for individual consumers. Because of the intermittent availability of sunlight and wind, they cannot by themselves supply all of our energy, but they can vastly reduce the amount of energy required from power plants. Installing 250,000 wind turbines could reduce the use of coal in the United States by two-thirds.[6]

Biofuels have become a much-discussed option for reducing carbon emissions. The idea is simple. We can produce fuels such as ethanol and diesel from plants, and burn them instead of fossil fuels. Burning biofuels puts carbon dioxide into the air, but it is carbon that plants have recently removed from the air, unlike the carbon from fossil fuels that plants took out of the air millions of years ago. Raising and producing the biofuels requires energy, particularly for those made from crops that are grown on modern farms that consume a lot of fuel (chapter 6). But even the least-efficient biofuels appear to produce more energy than they consume, resulting in a net reduction in carbon emissions (table 12.1).[7]

These calculations of the carbon benefits of biofuels do not, however, include the destruction of natural habitats. Although some of the

TABLE 12.1. Carbon Emissions Avoided by Different Practices over the Course of Thirty Years

PRACTICE	AVOIDED EMISSIONS (METRIC TONS OF CARBON PER HECTARE)
Sugarcane to ethanol	56
Wheat to ethanol	11
Sugar beet to ethanol	33
Maize to ethanol	12
Mustard seed to diesel	12
Woody biomass to diesel	57
Natural regrowth of tropical forest from crops	180
Conversion of crops to temperate pine forest	96
Conversion of crops to grassland	30

SOURCE: Renton Righelato and Dominick V. Spracklen, "Carbon Mitigation by Biofuels or by Saving and Restoring Forests?" *Science* 317 (2007): 902.

NOTE: 1 metric ton = 1.1 English tons; 1 hectare = 2.5 acres. The estimates of ethanol and diesel production do not include the carbon emissions that would result from the destruction of natural habitats in order to grow the crops.

plants from which biofuels are made can be grown in degraded land, most of the plant material would be grown in existing or newly created farmland. Once the carbon emissions from the destruction of natural habitats are taken into account, biofuel production may even release more carbon than it saves.[8] It is far better to produce biofuels from wastes, such as wood chips, than from materials that are grown specifically for biofuel production. This would also reduce the pressure that production of biofuels is placing on food prices. Biofuels derived from the managed and sustainable harvest of plant matter from natural habitats can even be carbon-negative, meaning that we can burn them while having a net removal of carbon dioxide from the atmosphere.[9] Biofuels can be used in many vehicles. In addition, all-electric or partially electric-powered cars are already available.[10]

An even better way to reduce carbon emissions is to restore natural habitats that have been damaged and to stop destroying the plant cover that we currently have. Forests have been a major part of the landscape for almost four hundred million years, and planting new forests merely replaces what has been destroyed, largely by human activity (chapter 1). Even without taking habitat destruction into account, restoring forests is still a much better way of reducing net carbon emissions than is the use of biofuels (table 12.1).

Burning coal, on the other hand, is one of the worst things we can do. Utility companies burn coal in power plants because it remains abundant and cheap. Coal, however, produces more carbon dioxide per unit of energy than any other source. Some utility companies, such as Duke Energy, have committed themselves to reducing carbon emissions. Others, like TXU Energy in Texas, did not. Utility companies are used to getting their own way. In 2006, TXU proposed the building of eleven coal-fired power plants in Texas, and Governor Rick Perry mandated that these facilities would not have to undergo the full range of environmental scrutiny. A federal judge found Perry's action unconstitutional in 2007, right before another company began the process of buying TXU and scrapping the plans for most of the coal-burning plants.[11]

A Brief Recent History of Resistance to Environmental Action

One thing we cannot do is to wait for our elected officials to solve the problems for us. To demonstrate this, I will outline the recent political history of environmental issues, especially of global warming. For six years, Congress was nearly inactive on global warming and related issues, except for a few lone voices such as those of Senators John McCain and Joseph Lieberman. The former chairman of the Senate Environment and Public Works Committee, James Inhofe of Oklahoma, is one of the last educated people in the world to believe that global warming is a hoax. Inhofe, who receives large contributions from the petroleum industry, was able to find an expert witness to testify that global warming is a hoax in a committee hearing in 2005: the science fiction writer Michael Crichton, whose novel *State of Fear* showed sinister environmental scientists placing bombs along Antarctic ice sheets to make them fall into the ocean. Even very few of his fellow Republicans in Washington, not to mention at state and local levels, took Inhofe seriously. Still, right up to his final days as head of the committee in December 2006, Inhofe's opinions remained vociferously unchanged. In such a climate of hostility, no progress could be made.

The hostility of the legislative branch was completely in accord with the policy of the administration of George W. Bush, which opposed all realistic action on these issues. The political appointees that headed the U.S. Environmental Protection Agency refused to recognize carbon dioxide as a pollutant over which they had jurisdiction. According to a national news report, the U.S. Department of Energy was required by law to issue guidelines for energy efficiency of consumer products; by March 2007 it had missed all thirty-four deadlines, and its own spokesperson admitted that the department had an abysmal record of promoting energy efficiency.

Appointed officials in the Bush administration made their loyalty visible in suppressing the scientific evidence of global warming and endangered species. In three cases, their successes were short-lived. The chief of staff at the White House Environmental Office, Philip Cooney, became infamous for altering government documents and Web sites, a practice that led to his resignation in 2005. Before working for the White House, he had been in charge of disseminating anti–global warming statements from the American Petroleum Institute; and when he resigned from the White House, he immediately went to work for ExxonMobil, one of the corporations that still opposed the science that documented global warming.[12] Meanwhile at NASA, George Deutsch, a writer in the public affairs office, tried to restrict climate scientist James Hansen from speaking about the scientific evidence of global warming. In 2006, Deutsch resigned when it was revealed that he had presented inaccurate credentials about his professional qualifications for his job.[13] Julie MacDonald, a deputy assistant secretary in the Department of the Interior, was supposed to help the Fish and Wildlife Service (FWS) fulfill its legal obligations of investigating the status of potentially endangered species. Instead she rewrote scientific reports, applied direct pressure on FWS employees to not report information that would show species to be endangered, and blocked the listing of endangered species and critical habitats. When her actions were revealed by an Inspector General's investigation in May 2007, MacDonald resigned.[14] In addition, scientists who worked for the Environmental Protection Agency (EPA) also felt pressure from the political appointees that headed the agency to present their results in a way that was compatible with Bush administration policy. In 2008, 889 of 1,586 respondents to an online survey conducted by the Union of Concerned Scientists indicated that they had experienced this pressure.[15]

U.S. government agencies often did not fulfill their legal obligation to enforce federal policies related to energy resources. For example, the Department of the Interior has a program within its Minerals Management Service to collect revenues from oil and gas companies that drill on federal land. The department's Inspector General Earl E. Devaney found in 2007 that the federal government had simply not collected very much of this revenue and that auditors who discovered the oversights had been blocked from recovering them.[16]

The pressure to resist the conclusions of global warming science came from the very top. President George W. Bush made a campaign promise in 2000 to place limits on carbon emissions; almost immediately after he was sworn into office, he announced that he had changed his mind. The governments of most nations have assented to the Kyoto Protocol, which is an international agreement about targets for reducing carbon dioxide emissions, launched in Kyoto, Japan, in 1997. The only major nations to not consent to this treaty were the United States and Australia. President Bush's opposition to the scientific consensus on global warming became so notorious that he finally had to make a conciliatory statement. In his 2006 State of the Union address, the president announced that America is addicted to oil, and that we need to do something about it. Part of the solution, he said, was switchgrass, a wild grass that can be fermented into fuel-grade ethanol. The very next day, botanists all over the United States, including me, received phone calls from local media asking about whether there were big fields of switchgrass just waiting to be mowed and fermented. Of course we had to inform the media that it would be possible to raise wild plants such as switchgrass for ethanol, but it would require many years and extensive investment. Meanwhile, the Bush administration slashed the budgets of the very agencies, such as the National Renewable Energy Laboratory, that would be able to spearhead such a project.[17] When attention was called to this seemingly hypocritical act, the administration claimed it was an oversight. The president refused to seek carbon emission standards. Instead, the Bush administration sought to reduce "greenhouse intensity," that is, the amount of carbon dioxide emission per unit of economic growth. Its proposal, however, allowed carbon emissions to continue at their present rate and even to increase.[18]

State and local governments in the United States did not wait for federal leadership on these important issues. Massachusetts sued the EPA in 2005 over its refusal to regulate carbon dioxide.[19] Several state governments, such as New York and California, went ahead and set their own carbon emission standards. As of June 2007, fourteen states had imposed carbon emission restrictions. Los Angeles was just one of many cities to adopt a "green plan" for economic development.[20]

Some corporations took action as well. An exception was ExxonMobil, which had a history of denying that climate change is a significant problem and spending money to suppress the truth about global warming. Meanwhile, other oil companies embraced the future and began to prepare for it. These corporations have joined with many others to form the Climate Action Partnership (CAP), which has thirty-three listed members as of February 2008.[21] These include three oil companies (the American branch of BP Amoco, the American branch of Royal Dutch Shell, and ConocoPhillips), Ford Motor Company, General Electric, Alcoa, DuPont, Caterpillar, Duke Energy, and IBM. CAP members want the federal government to establish carbon restrictions sooner rather than later so that they can go ahead and begin making their plans. The investment firms behind major corporations (such as Merrill Lynch) and insurance companies (as well as the reinsurers that insure the insurance companies) are worried about the effects of climate change on long-term business, and they also urge carbon emission limits.[22] Finally, in 2007, ExxonMobil announced that it was no longer denying the reality of global warming.[23]

Consumers continued to support corporations that seek energy efficiency. Without any leadership from Washington, the sales of CF light bulbs have skyrocketed, and there are waiting lines for the purchase of fuel-efficient vehicles. A great deal of resistance to global warming science came from the evangelical Christian Right. This continues to be the case with groups such as the Southern Baptists; however, many other Christian groups are working for the reduction of greenhouse emissions.[24]

Scientific and environmental organizations are sometimes accused of acting in their own financial interests (increasing their membership, selling books) by promoting the science of global warming. The profits of such organizations, however, are minuscule in comparison with the profits of corporations such as ExxonMobil, whose 2006, 2007, and 2008 profits were greater than those of any other corporation in history. The Competitive Enterprise Institute is an American antienvironmental "think tank" that received more than $2 million from ExxonMobil between 1998 and 2006, when ExxonMobil discontinued its support. The company gave almost $3 million to thirty-nine different antienvironmental organizations in 2005. As recently as February 2007, the American Enterprise Institute, funded largely by ExxonMobil, offered

$10,000 to any scientist who would dispute the findings of the International Panel on Climate Change regarding the human cause of global warming (see chapter 3).[25] This was widely viewed as unethical, as scientists are usually funded to do research rather than to decide in advance that they will reach a certain conclusion.

The Competitive Enterprise Institute published a book, *Global Warming and Other Eco-Myths*, in which climatologist John Christy claimed that global warming is not a certainty. His first graph was misleading. The data in the graph showed a clear upward trend of global temperatures since 1850. However, he broke the graph into several short arrows, some of which showed an increase and some of which did not. From this statistically deceptive procedure he concluded that temperatures may not be increasing.[26] This same climatologist said that if he had to choose between a world that is safe for the survival of whales and one that is safe for children, he would choose the latter—as if a world in which whales die is one that is safer for children.[27] How could an award-winning scientist make such errors? My interpretation is that political and religious beliefs blinded his judgment.

Another organization that vigorously attacked the science of global warming was the Heartland Institute. It sponsored a high-profile conference in New York City in March 2008 to discredit the scientific conclusions about global warming. It also published an advertisement that claimed nineteen thousand scientists had signed a petition questioning the validity of the human contribution to global warming. Commentators have cited this "list of thousands of scientists" on national radio without examining it. The list was posted on the Heartland Institute Web site. I looked up about a hundred of the names on the list. The credentials of the signers were not indicated, except for Ph.D. after some of the names. When I could locate information on the World Wide Web about the Ph.D. scientists, I could not find any climatologists among them. Their degrees were in such areas as medicine; industrial or agricultural chemistry; mechanical, civil, or electrical engineering; and theoretical physics. The Web site also claims that air pollution, asbestos, dioxin, lead, mercury, pesticides, and other pollutants pose no public health risk, and that all of the research that links smoking to negative health effects is "junk science." The Heartland Institute does not disclose its funding sources other than to state that no single corporation sponsors more than 5 percent of its budget. It appears to be nothing more than a clearinghouse for a whole range of viewpoints that have been soundly disproven. This is just one example of an organization that can mislead people who do not closely examine its claims.[28]

Then in 2007, after the inauguration of America's 110th Congress, nearly everything began to change. When the House of Representatives passed the Clean Long-Term Energy Alternative for the Nation (CLEAN) Act on January 18, 2007, during its first one hundred hours in session, it was a long overdue and much welcome move.[29] The House formed a new panel on energy independence and global warming, with Ed Markey (D-MA), as its chairman. A crop of bills sprouted, each a little different, but each seeking carbon emission limitations. Unfortunately, partisan politics quickly began to nullify the beginnings of congressional action on this issue. It took until December 2007 for Congress to pass an energy bill that required new vehicles to have an average fuel efficiency of 35 miles per gallon by 2020. This increased fuel efficiency may save 1.1 million barrels of oil per day—about half of what the United States currently imports from the Middle East.

Meanwhile, there was progress on the judicial front. In September 2007, a federal judge ruled that states have the legal right to set higher standards for emission control than the federal government does.[30] Despite this, the EPA refuses to allow California to set stricter carbon emission standards than those of the federal government. The Supreme Court has also ruled that the EPA has the power to regulate carbon dioxide as a pollutant. Because the leadership of the EPA declined to do this, attorneys general of eighteen states sued the agency in April 2008.

The world was closing in around the Bush administration. The International Panel on Climate Change (IPCC) issued reports in February, March, and April 2007 that presented a strong international scientific consensus that humans were major contributors to global warming, that the consequences would be severe, and that realistic remedial steps were possible (chapter 3). The IPCC published an overall summary in November 2007, in which it indicated not only that global warming and its consequences would be dramatic, but that they would be largely inevitable. Even if carbon emissions halted now, global warming and its consequences, such as water shortages and rising ocean levels, would continue for centuries because of the carbon dioxide emitted into the atmosphere during the twentieth century.[31] The other leaders of the G8 (the group of eight major economic powers in the world), especially German chancellor Angela Merkel, placed pressure on President Bush to accept carbon emission limitations, as other G8 nations had done. Britain has become the first nation to propose nationwide legally binding limits to carbon emissions.[32] In May 2007, with less than two weeks to go before the G8 Summit meeting, President Bush decided to reinvent himself as a leader in dealing with global warming by making his own carbon emission proposals. The very same week, in September 2007, in which the United Nations held a summit on global climate change, the president convened an alternative summit to promote voluntary carbon reductions. Critics have pointed out that Bush's proposals lacked sufficient enforcement. But it was clear that the antienvironmental stranglehold that characterized American politics from 2001 to 2006 was broken, at all levels of the federal government.

The momentum for change continued on the local level all around the world. For years, cities had been making their municipal buildings more environmentally responsible, primarily through energy conservation (chapter 4). Although cities cover only 1 percent of the earth's surface, they generate a large share of the greenhouse gases. Their actions to reduce energy use are therefore particularly important. In 2007, the William J. Clinton Foundation spearheaded a program, the Clinton Climate Initiative, which made loans totaling $1 billion available from major banks for cities to improve the energy efficiency of their buildings. The sixteen cities are New York, Houston, Toronto, Chicago, Mexico City, London, Berlin, Tokyo, Rome, Delhi, Karachi, Seoul, Bangkok, Melbourne, São Paolo, and Johannesburg. The cities expect to pay back their loans from their energy savings.[33] The World Bank was also ready to make investments in developing countries such as Papua New Guinea, Costa Rica, Indonesia, Brazil, and the Democratic

Republic of Congo in return for reductions in deforestation.[34] In all of these cases, saving forests and energy are seen not as alternatives to but as an essential part of economic preparedness for the future.

Environmental organizations continued to provide an outlet for people to not only promote carbon reductions, but to demand that the U.S. government take action. One example is the Energy Action Coalition, which brought hundreds of youth together in November 2007 to lobby the federal government for changes in energy policy.[35] And the American people were ready for change. According to a New York Times–CBS News poll published in April 2007, most Americans (84 percent) accepted the validity of a human contribution to global warming, and 90 percent of Democrats, along with 69 percent of Republicans, said immediate action was needed to curb carbon emissions. (Meanwhile, the fact that most Republican members of Congress still reject a restriction on carbon emissions shows that they are disconnected from their own party's constituents.) If a choice was necessary between protecting the environment and stimulating the economy, 52 percent chose the environment, and only 36 percent chose the economy. While 21 percent of respondents indicated that we needed to increase petroleum production, more than three times as many (68 percent) said that we should put greater emphasis on energy conservation. Fully 92 percent favored increased fuel efficiency standards for vehicles. Three-quarters were willing to pay more for electricity in order to promote the use of renewable energy sources by utilities.[36] A poll published by the Yale Project on Climate Change in October 2007 showed similar patterns. This survey indicated that nearly half of Americans think global warming is having, or will soon have, dangerous impacts on people around the world—a twenty-point increase from 2004 poll results. More than two-thirds of the respondents favored carbon emissions reductions even greater than those required by the Kyoto Protocol. More than 80 percent favored a mandatory increase in fuel efficiency in vehicles, and the use of renewable energy by utilities, even if it resulted in automobiles and annual utility bills that would be up to several hundred dollars more expensive.

Change continued around the world in 2007. A national election in the fall of that year changed the leadership of Australia, which is now prepared to cooperate with other nations to reduce carbon emissions. In December 2007, a conference in Bali sponsored by the United Nations laid the groundwork for a new international agreement on carbon emissions, intended to replace the Kyoto Protocol when it expires in 2012. This time, the United States signed on to the agreement.

A Time for Passion

The collapse of U.S. federal opposition to global warming science does not mean that the government will begin to solve the problems. As always, it will be the decisions and passions of individuals that make the difference. Just as no simple government policy can solve the problems of global warming, deforestation, and soil erosion, the easy actions that citizens can take are insufficient to solve these problems.

An increasing number of people choose the easy way to reduce their impact on the world, especially in terms of carbon emissions: they purchase carbon offsets. That is, they go ahead and produce as much carbon dioxide as they like but pay someone else to produce less or to plant trees. Sometimes this is a very effective thing to do. More than a decade ago, the state of California required utility companies to reduce their air pollution, but the utilities countered that it would cost many billions of dollars to do so. Instead, they proposed to reduce the total amount of air pollution in the state by an equivalent amount, but by a different means: they purchased new low-emissions cars for poor families that drove old, fuming cars—a compromise the state found acceptable.

But often, the purchase of carbon offsets is nothing more than a way of assuaging guilt. Delta Air Lines allows online ticket purchasers to fund tree plantings to compensate for the emissions resulting from air travel. Or, if driving 12,000 miles in your vehicle produces 20,000 pounds of carbon, you can send $80 to TerraPass, a company that specializes in selling carbon offsets. And the Vatican can now go ahead and generate as much carbon as it would like, because a Hungarian company has offered, as a charitable contribution, to plant enough trees to compensate for the Vatican's emissions.[37]

Carbon offsets are better than doing nothing. At least in the examples cited above, the offset is proportional to the amount of carbon generated by the purchaser. In contrast, the image of a large church congregation planting a tree for Earth Day, next to a parking lot full of gas-guzzling vehicles, is laughable. But carbon offsets create the false illusion that, if you buy them, you have discharged your duty to the human race. The Transnational Institute's Carbon Trade Watch called the carbon offsets associated with occasional tree planting as "the sale of offset indulgences," making reference to the medieval practice of paying priests for the forgiveness of sins rather than undertaking the hard work of living a good life.[38] It is a carbon dioxide version of what theologian Dietrich Bonhoeffer called "cheap grace." Author Peter Schweitzer takes this idea even further: if we can buy a carbon offset to compensate for our excessive consumption, why can't we buy other kinds of offsets? He suggests that you could buy an adultery offset by donating to a pro-family organization, or a tofu offset by giving money to a vegetarian; these actions would allow you to go ahead with your compromising and gluttonous lifestyle with a clear conscience.[39]

Carbon offsets are an incomplete solution because they may not actually reduce the amount of carbon emissions produced by the purchaser, or perhaps even in the world. If an organization pays a developer to not cut down a forest, the trees will be saved, and both parties feel that they benefit. But both the developer and the donors should, in fact, save trees and reduce their carbon emissions anyway. I drive my Toyota even though nobody is paying me an offset. The world would be no better off if someone paid me; in fact, it could be worse, because the donors might continue to drive vehicles that produce a lot of emissions.

There is no one, single thing we can do, but the many small things we can do may add up to significant progress. Americans can save a lot of gasoline by planning their trips around town better—doing more things on fewer trips. This, of course, also saves time and money. None of these things—planting trees, paying someone else to plant trees, installing a few CF bulbs, driving a small car, driving more efficiently—is enough, but they add up. Small things, like sealing up the gaps around windows through which heat is lost in winter, or turning the thermostat down in winter or up in summer by a couple of degrees, can add up to enough energy savings to completely compensate for all of the oil that could be delivered from drilling the Arctic National Wildlife Refuge (ANWR). The savings resulting from energy efficiency are available immediately (unlike the decade that would be required for developing the ANWR field, especially as the permafrost melts under the equipment and roads) and for very little (and sometimes no) cost. When I save energy by walking rather than driving, my health improves (except when I have to breathe the fumes of vehicles driving past me).

Coming up with a complex list of things to do may sound overwhelming. And if each of these things were a disconnected item, it might seem that the only way we can save the earth from ecological disaster is to spend all of our time thinking of lots of little ways to do it, perhaps by writing the long, long list on the back of previously used paper, which we can later recycle. But there is good news here also. They are not disconnected items. They can result from a simple shift in attitude. All that is necessary is that you start thinking about what you are doing. Look around the room. Are there simple things you can do to reduce the electricity costs of lighting and air conditioning? Maybe closing the drapes in the afternoon on south-facing windows? Or using a fan rather than an air conditioner except on the hottest days? Think ahead in your plans. Is there some way you can enjoy a vacation closer to home, or at home, thus reducing the expenditure of gasoline? Each person can generate his or her own list. It becomes enjoyable and creative, rather than a burden. You do not need a book titled "One Hundred Thousand Things You Can Do to Save the Planet." And, if my experience has any relevance to what yours is or would be, you will feel good about doing something positive. And you will feel a little less trapped by high fuel costs.

But this shift in attitude has to be a large and permanent one. Doing just a couple of things, or doing them for only a week, will have little effect on the overall human impact on the earth. Of course it is important for all of us to use less energy and use less stuff. But we also need to *enjoy* using less energy and stuff. Recycling is an example. Recycling is such a part of my life now that I hardly think about it, and it certainly does not feel like a burden. I am content with a life that has a low impact on the earth, compared with that of the average American; and millions of other Americans can say the same thing about their lives as well.

It is a life that can be described by the word *frugality*. The American mythology of unlimited capitalistic growth has made this a dirty word; buying and consuming as much as possible is said to be good for the economy. But those of us who live frugally can testify that there is great happiness and contentment to be found in a lifestyle that is less wasteful. The merging of frugality and contentment is depicted in the village visited by the wanderer in Akira Kurosawa's classic film *Dreams*. The village has no electricity. The wanderer asks an old man whether it bothers him that they have no electric lights at night. The old man tells him: "Night is supposed to be dark." The horrible environmental records of former communist countries demonstrate that capitalism, per se, is not the problem. Saving the world from problems such as the greenhouse effect requires not so much a new style of government as a change of viewpoint, an attitude of contentment with consuming less of the world's resources. Although much of our expenditure of energy and material is necessary to keep our lives from being unpleasant (think about laundering your clothes with a washboard and a bucket), perhaps an even greater amount of our expenditure is unnecessary. For example, it is good to have electric light at night. But much of the electric light in our cities is wasted, going straight up into outer space rather than down on the street where it can protect us from crime.

Our shift in attitude needs to grow out of a sense of mission. Minimizing our impact on the earth is not something we do instead of dealing with political and economic concerns. "Environmental issues" involve all of the things that we do to, with, and for all of the other people with which we share the earth. "The environment" is, in fact, *the medium through which we interact* with all of the people in the world. The global warming that I generate does not just make the weather a little hotter outside my house in the summer; it contributes to rising oceans that can obliterate entire island nations and bring droughts to Africa. Environmental issues have not traditionally been popular among conservative Christians (see above). But Norman Geisler, a conservative Christian theologian, pointed out more than a decade ago that if all men are brothers (using a traditional male-centered metaphor), we should care about the earth, "for it is my brother's Earth."[40]

A nation's positive and negative environmental actions have a major impact on world events. Topsoil erodes from the farmland of one nation and washes down the river through another nation, depriving the first of agricultural production and subjecting the second to floods and mudslides. If the second nation took an equivalent amount of topsoil by force, the first would take military action in response. Nations go to war over much less soil than we are currently losing as a result of deforestation. And it cannot escape anyone's notice that nations go to war to protect resources such as oil. As political scientist Kevin Phillips has pointed out, American and European involvement in the Middle East has been motivated by petroleum for at least a hundred years. The American insistence that the invasion of Iraq had nothing to do with oil is only the most recent of a long string of denials on the part of several Western nations.[41] The United States currently spends about $250 million per day on the Iraq War. Science writer Bill McKibben wrote that "gas sucking SUVs . . . should by all rights come with their own little Saudi flags on the hood."[42] As Phil Clapp, president of the National Environmental Trust, says, the entire world economy is built on a bet on how long the House of Saud can continue.[43] The U.S. government has identified energy independence as an essential part of national security but has yet to institute a realistic vision for achieving this.

Everything that we can do to reduce energy use and find sustainable and renewable energy sources is therefore an act of patriotism. I have recycled for a long time, but the rest of my family did not consistently do so until September 11, 2001, when they decided it was important to begin being *environmentally patriotic*. We cannot selfishly waste energy and throw everything away in a world that we share with humans whose lives are in jeopardy.

The United States leads the world in carbon dioxide emissions. The United States therefore bears a vastly disproportionate responsibility to reduce atmospheric carbon dioxide and has a vastly disproportionate effect on the world when it (and we) fail to do so. It cannot do so alone. The European Union has taken strong steps toward carbon restrictions, without American cooperation; but it is unlikely that nations such as China and India will do so unless the United States leads the way. As Wangari Maathai points out, the United States produces the most carbon emissions, but Africa will suffer the most from the droughts and storms that will result from global warming. We have an international responsibility to reduce our carbon emissions, to save our own trees, and to help other countries save their trees.[44] Maathai is not alone in this opinion. British Prime Minister Gordon Brown said in July 2007: "We know that the gains from global prosperity have been disproportionately enjoyed by the people in industrialized countries and that the consequences of climate change will be disproportionately felt by the poorest who are least responsible for it—making the issue of climate change one of justice as much as economic development."[45] If all the people in the less-developed countries were able to adopt American standards of living and produce as much carbon per capita as we do, world carbon emissions might increase by a factor of ten. They will not be able to do this, however, and they are resentful toward us that we are so wasteful.

Finally, we will have to go beyond national patriotism, even beyond human love, to solve the problems on this planet. It is not necessary to go as far as Francis Hopkinson, a signer of the American Declaration of Independence, who said in 1782: "Trees, as well as men, are capable of enjoying the rights of citizenship and therefore ought to be protected . . . ," but we have to recognize plants as legitimate sharers of this planet with us.[46] If we give trees and other plants the respect that is necessary for them to continue to do their jobs, we will enjoy two benefits. First, the act of respect will make us reduce our environmentally destructive actions to a level that the plants, masters of the earth's ecosystems, can handle. Second, once this occurs, the plants will help to save us from our environmental problems. Plants are not just crucial for the economic value that they produce when harvested, but a million times more important for what they do. To say that a tree is worth only its lumber is even worse than saying a bear is worth only the rug that can be made from its skin.

From the most ancient times, the collapse of civilizations has been associated with, and largely caused by, environmental degradation. The people who experienced these events did not understand what caused them and frequently attributed the crisis to the disapproval of the gods. This was the explanation provided in *The Curse of Akkad* (chapter 1). It was also the case in the ancient kingdoms of Israel and Judah, as interpreted by the writers of the Old Testament books of the Chronicles. When the kingdom of Judah was at last conquered by Babylon, the chronicler attributed it to the sins of the people. But there is one particular sin that the writer singled out. One of the many Old Testament commandments was to allow agricultural land to lie fallow every seventh year. This "sabbath of the fields," we now understand, would have allowed wild plants to partially restore the fertility of the soil and diseases and pests specific to the crops to die away. To the ancient Israelites, it was perhaps solely understood in terms of religious service. But the sabbath of the fields was never put into practice. Thus, the chronicler implied, the people of Judah owed God one-seventh of the years that their kingdom had existed. According to tradition, Judah existed for 490 years, so the people therefore owed the land and its God 70 years of rest. They did not render this offering while the kingdom stood, so the land rested during the seventy years of Israelite captivity in Babylon. Among the very last words of the second book of Chronicles is this chilling statement: "And so the land enjoyed its rests. All the days that it lay desolate it kept Sabbath, to fulfill seventy years." The implication, as much for us as for those ancient people, is that one way or another the land will get its rest, its chance to recover. We do not need to stop our economic activities in order to allow this to happen. We can incorporate a modern equivalent of the sabbath of the fields into our economic system by replanting forests or by using natural systems agriculture. In this way, the land can rest the way the heart rests—between each contraction, without missing a beat. But our choice is clear: we either incorporate stewardship of the land into our economic system, allowing plants to renew the land on a regular basis, or else the land will enjoy its sabbaths after the collapse of our economic system. Now, or later, the land will rest.

Imagine a machine that can solve all of our problems. It produces oxygen for us to breathe, food for us to eat, medicines and many other products we need in our lives, and protects us from floods, droughts, and overheating by removing carbon dioxide from the air. To build and operate a billion such machines would produce more carbon dioxide than the machines would remove from the air, unless we could get the machines to build themselves and run on solar energy. Plants do all of these things. We do not have to make them do it, or even to thank them, but just allow them. And they do it in complete silence and unutterable beauty.

We can, therefore, do little better than what Adlai Stevenson told us in his last speech in 1965, before the widespread modern recognition of our ecological crises: "We travel together, passengers in a little space-ship . . . preserved from annihilation only by the care, the work, and, I will say, the love we give our fragile craft."[47]

notes

Introduction

1. James A. Schrader, William R. Graves, Stanley A. Rice, and J. Phil Gibson, "Differences in Shade Tolerance Help Explain Varying Success in Two Sympatric *Alnus* Species," *International Journal of Plant Sciences* 167 (2006): 979–989. J. Phil Gibson, Stanley A. Rice, and Clare M. Stucke. "Comparison of population genetic diversity between a rare, narrowly distributed species and a common, widespread species of *Alnus* (Betulaceae)," *American Journal of Botany* 95 (2008): 588–596.
2. David Beerling, *The Emerald Planet: How Plants Changed Earth's History* (Oxford: Oxford University Press, 2007), 3.
3. Robert Costanza et al., "The Value of the World's Ecosystem Services and Natural Capital," *Nature* 387 (1997): 253–260.

Chapter 1. An Injured Paradise

Epigraph: W. S. Merwin, *Migration: New and Selected Poems* (Port Townsend, Wash.: Copper Canyon Press, 2007).
1. E. C. Pielou, *After the Ice Age: Return of Life to Glaciated North America* (Chicago: University of Chicago Press, 1992). Thomas M. Bonnicksen, *America's Ancient Forests: From the Ice Age to the Age of Discovery* (New York: Wiley, 2000).
2. Brian Fagan, *The Long Summer: How Climate Changed Civilization* (New York: Basic Books, 2004).
3. Anna Lewington and Edward Parker, *Ancient Trees: Trees That Live for a Thousand Years* (London: Collins and Brown, 1999). Thomas Pakenham, *Remarkable Trees of the World* (New York: Norton, 2002).
4. John Perlin, *A Forest Journey: The Role of Wood in the Development of Civilization* (Cambridge, Mass.: Harvard University Press, 1989). Patricia L. Fall et al., "Environmental Impacts of the Rise of Civilization in the Southern Levant," chap. 7 in Charles L. Redman, Steven R. James, Paul R. Fish, and J. Daniel Rogers, eds., *The Archaeology of Global Change: The Impact of Humans on Their Environment* (Washington, D.C.: Smithsonian Institution Press, 2004).
5. Anthony D. Barnosky, "Assessing the Causes of Late Pleistocene Extinctions on the Continents," *Science* 306 (2004): 70–75.
6. Charles C. Mann, *1491: New Revelations of the Americas before Columbus* (New York: Knopf, 2005).
7. William Bartram, *Travels through North and South Carolina, Georgia, East and West Florida, the Cherokee Country, the Extensive Territories of the Muscogulges or Creek Confederacy, and the Country of the Chactaws* (1791; repr., Savannah, Ga.: Beehive Press, 1973).

8. Mann, *1491*.
9. Michael Balter, "Seeking Agriculture's Ancient Roots," *Science* 316 (2007): 1830–1835. Ehud Weiss, Mordechai E. Kislev, and Anat Hartmann, "Autonomous Cultivation before Domestication," *Science* 312 (2006): 1608–1610. David Rindos, *The Origins of Agriculture: An Evolutionary Perspective* (San Diego: Academic Press, 1984).
10. Jared Diamond, *Guns, Germs, and Steel: The Fates of Human Societies* (New York: Norton, 1997).
11. Karl-Ernst Behre, "The Role of Man in European Vegetation History," in B. Huntley and T. Webb III, eds., *Vegetation History*, 633–672 (Dordrecht, Netherlands: Kluwer Scientific Publications, 1988).
12. Mann, *1491*.
13. Bartram, *Travels*.
14. Michael Tennesen, "Black Gold of the Amazon," *Discover*, April 2007, 46–52.
15. Bartram, *Travels*.
16. Suzanne K. Fish and Paul R. Fish, "Unsuspected Magnitudes: Expanding the Scale of Hohokam Agriculture," chap. 11 in Redman et al., *Archaeology of Global Change*.
17. Jeffrey S. Dean, "Anthropogenic Environmental Change in the Southwest as Viewed from the Colorado Plateau," chap. 10 in Redman et al., *Archaeology of Global Change*.
18. Charles L. Redman, "Environmental Degradation and Early Mesopotamian Civilization," chap. 8 in Redman et al., *Archaeology of Global Change*.
19. Elizabeth Kolbert, "The Curse of Akkad," chap. 5 in *Field Notes from a Catastrophe: Man, Nature, and Climate Change* (New York: Bloomsbury, 2006).
20. Naomi F. Miller, "Long-Term Vegetation Changes in the Near East," chap. 6 in Redman et al., *Archaeology of Global Change*.
21. Perlin, *Forest Journey*.
22. Sander E. Van der Leeuw, François Favory, and Jean-Jacques Girardot, "The Archaeo-logical Study of Environmental Degradation: An Example from Southwestern France," chap. 5 in Redman et al., *Archaeology of Global Change*.
23. Jared Diamond, *Collapse: How Societies Choose to Fail or Succeed* (New York: Viking, 2005).
24. Sarah L. O'Hara and Sarah E. Metcalfe, "Late Holocene Environmental Change in West-Central Mexico," chap. 4 in Redman et al., *Archaeology of Global Change*.
25. Diamond, *Collapse*.
26. Timothy A. Kohler, "Pre-Hispanic Human Impact on Upland North American South-western Environments: Evolutionary Ecological Perspectives," chap. 12 in Redman et al., *Archaeology of Global Change*.
27. Roland Bechmann, *Trees and Man: The Forest in the Middle Ages*, trans. Katharyn Dunham (New York: Paragon House, 1990).
28. Conrad Totman, *The Green Archipelago: Forestry in Pre-industrial Japan* (Athens: University of Ohio Press, 1998).
29. Alfred W. Crosby, *Ecological Imperialism: The Ecological Expansion of Europe, 900–1900* (New York: Cambridge University Press, 1986).
30. Richard V. Francaviglia, *The Cast Iron Forest: A Natural and Cultural History of the North American Cross Timbers* (Austin: University of Texas Press, 2000).
31. K. J. Willis, L. Gillson, and T. M. Brncic, "How 'Virgin' Is Virgin Rainforest?" *Science* 304 (2004): 402–403.
32. Susanna Hecht and Alexander Cockburn, *The Fate of the Forest: Developers, Destroyers, and Defenders of the Amazon* (New York: HarperCollins, 1989). Peter Dauvergne, *Shadows in the Forest: Japan and the Politics of Timber in Southeast Asia* (Cambridge, Mass.: MIT Press, 1997).

33. Marc D. Abrams, "Red Maple Paradox," *BioScience* 48 (1998): 1–18.
34. Charles E. Little, *The Dying of the Trees: The Pandemic in America's Forests* (New York: Penguin, 1995), 19.
35. John Terborgh, *Requiem for Nature* (Covelo, Calif.: Island Press, 1999).
36. Peter D. Ward and Alexis Rockman, *Future Evolution: An Illuminated History of Life to Come* (New York: Henry Holt, 2001).
37. Wangari Maathai, *The Green Belt Movement: Sharing the Approach and the Experience* (New York: Lantern Books, 2003).
38. David Brower, with Steve Chapple, *Let the Mountains Talk, Let the Rivers Run: A Call to Those Who Would Save the Earth* (New York: HarperCollins, 1995).

Chapter 2. Plants Put the Oxygen in the Air

1. David Beerling, *The Emerald Planet: How Plants Changed Earth's History* (Oxford: Oxford University Press, 2007).
2. Christopher B. Field, Michael J. Behrenfeld, James T. Randerson, and Paul Falkowski, "Primary Production of the Biosphere: Integrating Terrestrial and Oceanic Components," *Science* 281 (1998): 237–240.
3. Michael J. Benton, *When Life Nearly Died: The Greatest Mass Extinction of All Time* (London: Thames and Hudson, 2003). Douglas H. Erwin, *Extinction: How Life on Earth Nearly Ended 250 Million Years Ago* (Princeton, N.J.: Princeton University Press, 2006). Beerling, *Emerald Planet*.
4. Raymond B. Huey and Peter D. Ward, "Hypoxia, Global Warming, and Terrestrial Late Permian Extinctions," *Science* 308 (2005): 398–401. Peter D. Ward et al., "Abrupt and Gradual Extinction among Late Permian Land Vertebrates in the Karoo Basin, South Africa," *Science* 307 (2005): 709–714.
5. John Bohannon, "Microbe May Push Photosynthesis into Deep Water," *Science* 308 (2005): 1855.

Chapter 3. Greenhouse Earth

1. Intergovernmental Panel on Climate Change, *Climate Change 2007: The Physical Science Basis. Summary for Policymakers*, available at http://www.ipcc.ch/SPM2feb07.pdf (accessed September 12, 2007). Qiang Fu et al., "Enhanced Mid-latitude Tropospheric Warming in Satellite Measurements," *Science* 312 (2006): 1179. National Oceanic and Atmospheric Administration, "A Paleo Perspective on Global Warming: The Instrumental Data of Past Global Temperatures," available at http://www.ncdc.noaa.gov/paleo/globalwarming/instrumental.html (accessed September 12, 2007).
2. Rowan T. Sutton and Daniel L. R. Hodson, "Atlantic Ocean Forcing of North American and European Summer Climate," *Science* 309 (2005): 115–118.
3. Richard A. Kerr, "Record U.S. Warmth of 2006 Was Part Natural, Part Greenhouse," *Science* 317 (2007): 182–183.
4. Center for the Study of Carbon Dioxide and Global Change, "CO_2 Science," available at http://www/co2science.org (accessed July 17, 2007).
5. Naomi Oreskes, "The Scientific Consensus on Climate Change," *Science* 306 (2004): 1686.

6. Michael E. Mann, R. S. Bradley, and M. K. Hughes, "Global-Scale Temperature Patterns and Climate Forcing over the Past Six Centuries," *Nature* 392 (1998): 779–787. Timothy J. Osborn and Keith R. Briffa, "The Spatial Extent of 20th-Century Warmth in the Context of the Past 1200 Years," *Science* 311 (2006): 841–844. National Oceanic and Atmospheric Administration, "A Paleo Perspective on Global Warming: Paleoclimatic Data for the Last 2000 Years," available at http://www.ncdc.noaa.gov/paleo/globalwarming/paleolast.html (accessed September 12, 2007). P. D. Jones and M. E. Mann, "Climate over Past Millennia," *Reviews of Geophysics* 42 (2004): RG2002, doi:10.1029/2003RG000143.

7. Geoff Brumfiel, "Academy Affirms Hockey-Stick Graph," available at http://www.nature.com/news/2006/060626/full/4411032a.html (accessed September 12, 2007). RealClimate, "RealClimate: Climate Science from Climate Scientists," available at http://www.realclimate.org (accessed September 12, 2007).

8. International Panel on Climate Change, *Climate Change 2007: The Physical Science Basis.*

9. J. R. Petit et al., "Historical Isotopic Temperature Record from the Vostok Ice Core," in *Trends: A Compendium of Data on Global Change*, Carbon Dioxide Information Analysis Center, Oak Ridge National Laboratory, 2000; available at http://cdiac.ornl.gov/trends/temp/vostok/jouz_tem.htm (accessed April 21, 2006).

10. Robin E. Bell, "The Unquiet Ice," *Scientific American*, February 2008, 60–67. Anny Cazenave, "How Fast Are the Ice Sheets Melting?" *Science* 314 (2006): 1250–1252. J. L. Chen, C. R. Wilson, and B. D. Tapley, "Satellite Gravity Measurements Confirm Accelerated Melting of Greenland Ice Sheet," *Science* 313 (2006): 1958–1960. Ian M. Howat, Ian Joughin, and Ted A. Scambos, "Rapid Changes in Ice Discharge from Greenland Outlet Glaciers," *Science* 315 (2007): 1559–1561. Richard A. Kerr, "A Worrying Trend of Less Ice, Higher Seas," *Science* 311 (2006): 1698–1701. S. B. Luthcke et al., "Recent Greenland Ice Mass Loss by Drainage System from Satellite Gravity Observations," *Science* 314 (2006): 1286–1289. Mark F. Meier et al., "Glaciers Dominate Eustatic Sea-Level Rise in the 21st Century," *Science* 317 (2007): 1064–1067. Bette L. Otto-Bliesner et al., "Simulating Arctic Climate Warmth and Icefield Retreat in the Last Interglaciation," *Science* 311 (2006): 1751–1753. Jonathan T. Overpeck et al., "Paleoclimatic Evidence for Future Ice-Sheet Instability and Rapid Sea-Level Rise," *Science* 311 (2006): 1747–1750. Andrew Shepherd and Duncan Wingham, "Recent Sea-Level Contributions of the Antarctic and Greenland Ice Sheets," *Science* 315 (2007): 1529–1532. Martin Truffer and Mark Fahnestock, "Rethinking Ice Sheet Time Scales," *Science* 315 (2007): 1508–1510. David G. Vaughan and Robert Arthern, "Why Is It Hard to Predict the Future of Ice Sheets?" *Science* 315 (2007): 1503–1504. Isabella Velicogna and John Wahr, "Measurements of Time-Variable Gravity Show Mass Loss in Antarctica," *Science* 311 (2006): 1754–1756.

11. Tim P. Barnett et al., "Penetration of Human-Induced Warming into the World's Oceans," *Science* 309 (2005): 284–287.

12. Richard A. Kerr, "Mother Nature Cools the Greenhouse, but Hotter Times Still Lie Ahead," *Science* 320 (2008): 595.

13. Richard Seager et al., "Model Projections of an Imminent Transition to a More Arid Climate in Southwestern North America," *Science* 316 (2007): 1181–1184. Tim P. Barnett et al., "Human-Induced Changes in the Hydrology of the Western United States," *Science* 319 (2008): 1080–1083. Robert Kunzig, "Drying of the West," *National Geographic*, February 2008, 90–113.

14. A. L. Westerling et al., "Warming and Earlier Spring Increase Western U.S. Forest Wildfire Activity," *Science* 313 (2006): 940–943; summary by Steven W. Running, "Is Global Warming Causing More, Larger Wildfires?" *Science* 313 (2006): 927–928.

15. C. D. Hoyos et al., "Deconvolution of the Factors Contributing to the Increase in Global Hurricane Intensity," *Science* 312 (2006): 94–97. Kevin E. Trenberth, "Warmer Oceans, Stronger Storms," *Scientific American*, July 2007, 44–47. P. J. Webster et al., "Changes in Tropical Cyclone Number, Duration, and Intensity in a Warming Environment," *Science* 309 (2005): 1844–1846; summary by Richard A. Kerr, "Is Katrina a Harbinger of Still More Powerful Hurricanes?" *Science* 309 (2005): 1807.

16. B. N. Goswami et al., "Increasing Trend of Extreme Rain Events over India in a Warming Environment," *Science* 314 (2006): 1442–1445.

17. Abraham J. Miller-Rushing et al., "Photographs and Herbarium Specimens as Tools to Document Phenological Changes in Response to Global Warming," *American Journal of Botany* 93 (2006): 1667–1674.

18. Xiaoyang Zhang et al., "Diverse Responses of Vegetation Phenology to a Warming Climate," *Geophysical Research Letters* 34 (2007); available at http://www.agu.org/pubs/crossref/2007/2007GL031447.shtml (accessed February 11, 2008).

19. Camille Parmesan and Gary Yohe, "A Globally Coherent Fingerprint of Climate Change Impacts across Natural Systems," *Nature* 421 (2003): 37–42. G. Grabherr, M. Gottfried, and H. Pauli, "Climate Effects on Mountain Plants," *Nature* 369 (1994): 448. R. B. Myneni et al., "Increased Plant Growth in the Northern High Latitudes from 1981 to 1991," *Nature* 386 (1997): 698–702; commentary by I. Fung, "A Greener North," *Nature* 386 (1997): 659–660.

20. Daniel W. McKenney et al., "Potential Impacts of Climate Change on the Distribution of North American Trees," *BioScience* 57 (2007): 939–948.

21. A. H. Fitter and R.S.R. Fitter, "Rapid Changes in Flowering Time in British Plants," *Science* 296 (2002): 1689–1691.

22. Mario Sanz-Elorza et al., "Changes in the High-Mountain Vegetation of the Central Iberian Peninsula as a Probable Sign of Global Warming," *Annals of Botany* 92 (2003): 273–280.

23. Chris D. Thomas and Jack J. Lennon, "Birds Extend Their Ranges Northward," *Nature* 399 (1999): 213. H.Q.P. Crick et al., "UK Birds Are Laying Eggs Earlier," *Nature* 388 (1997): 526. Anne Charmantier et al., "Adaptive Phenotypic Plasticity in Response to Climate Change in a Wild Bird Population," *Science* 320 (2008): 800–803.

24. Camille Parmesan et al., "Poleward Shifts in Geographical Ranges of Butterfly Species Associated with Regional Warming," *Nature* 399 (1999): 579–583.

25. Stuart Bearhop et al., "Assortative Mating as a Mechanism for Rapid Evolution of a Migratory Divide," *Science* 310 (2005): 502–504.

26. Çağan H. Şekercioğlu et al., "Climate Change, Elevational Range Shifts, and Bird Extinctions," *Conservation Biology* 22 (2008): 140–150.

27. Marcel E. Visser and Leonard J. M. Holleman, "Warmer Springs Disrupt the Synchrony of Oak and Winter Moth Phenology," *Proceedings of the Royal Society of London B (Biology)* 286 (2001): 289–294.

28. Marcel E. Visser, Leonard J. M. Holleman, and Phillip Gienapp, "Shifts in Caterpillar Biomass Phenology due to Climate Change and Its Impact on the Breeding Biology of an Insectivorous Bird," *Oecologia* 147 (2006): 164–172.

29. Daniel H. Nussey et al., "Selection on Heritable Phenotypic Plasticity in a Wild Bird Population," *Science* 310 (2005): 304–306.

30. Christine A. Rogers, Fakhri Bazzaz, et al., "Interaction of the Onset of Spring and Elevated Atmospheric CO_2 on Ragweed (*Ambrosia artemesiifolia* L.) Pollen Production," *Environmental Health Perspectives* 114 (2006): 865–869.

31. P. A. Umina et al., "A Rapid Shift in a Classic Clinal Pattern in *Drosophila* Reflecting Climate Change," *Science* 308 (2005): 691–693.
32. Stanley A. Rice, "Speciation," in *Encyclopedia of Evolution* (New York: Facts on File, 2006).
33. Richard A. Kerr, "Is Battered Arctic Sea Ice down for the Count?" *Science* 318 (2007): 33–34. Andrew C. Revkin, "Scientists Report Severe Retreat of Arctic Ice," *New York Times*, September 21, 2007.
34. K. B. Suttle, Meredith A. Thomsen, and Mary E. Power, "Species Interactions Reverse Grassland Responses to Changing Climate," *Science* 315 (2007): 640–642.
35. Fred Pearce, *With Speed and Violence: Why Scientists Fear Tipping Points in Climate Change* (Boston: Beacon, 2007).
36. Evan Mills, "Insurance in a Climate of Change," *Science* 309 (2005): 1040–1044.
37. CNA Corporation, "National Security and the Threat of Climate Change," available at http://securityandclimate.cna.org/report (accessed September 16, 2007). David B. Lobell et al., "Prioritizing Climate Change Adaptation Needs for Food Security in 2030," *Science* 319 (2008): 607–610; reviewed by Molly E. Brown and Christopher C. Funk, "Food Security under Climate Change," *Science* 319 (2008): 580–581.
38. Intergovernmental Panel on Climate Change, *Climate Change 2007: Climate Change Impacts, Adaptation, and Vulnerability. Summary for Policymakers*, available at http://www.ipcc.ch/SPM6avr07.pdf (accessed September 16, 2007).
39. Stefan Rahmsdorf et al., "Recent Climate Observations Compared to Projections," *Science* 316 (2007): 709.
40. Keith P. Shine and William T. Sturges, "CO_2 Is Not the Only Gas," *Science* 315 (2007): 1804–1805.
41. National Oceanic and Atmospheric Administration, "Trends in Carbon Dioxide," available at http://www.esrl.noaa.gov/gmd/ccgg/trends/co2_data_mlo.html (accessed September 16, 2007).
42. J.-M. Barnola et al., "Historical Carbon Dioxide Record from the Vostok Ice Core," in *Trends: A Compendium of Data on Global Change.* Carbon Dioxide Information Analysis Center, Oak Ridge National Laboratory, 2000; available at http://cdiac.ornl.gov/trends/co2/vostok.htm (accessed April 21, 2006).
43. Michael I. Mishchenko et al., "Long-Term Satellite Record Reveals Likely Recent Aerosol Trend," *Science* 315 (2007): 1543.
44. Intergovernmental Panel on Climate Change, *Climate Change 2007: The Physical Science Basis.* William Collins et al., "The Physical Science behind Climate Change," *Scientific American*, August 2007, 64–73.
45. Urs Siegenthaler et al., "Stable Carbon Cycle–Climate Relationship during the Late Pleistocene," *Science* 310 (2005): 1313–1317. Renato Spahni et al., "Atmospheric Methane and Nitrous Oxide of the Late Pleistocene from Antarctic Ice Cores," *Science* 310 (2005): 1317–1321.
46. Mark Pagani et al., "Marked Decline in Atmospheric Carbon Dioxide Concentrations during the Paleogene," *Science* 309 (2005): 600–603.
47. Mark Pagani et al., "An Ancient Carbon Mystery," *Science* 314 (2006): 1556–1557.
48. Isabel P. Montañez et al., "CO_2-Forced Climate and Vegetation Instability during Late Paleozoic Deglaciation," *Science* 315 (2007): 87–91. Gregory J. Retallack, "A 300-Million-Year Record of Atmospheric Carbon Dioxide from Fossil Cuticles," *Nature* 411 (2001): 287–290.
49. "Carbon emissions" refers to the weight of just the carbon that is emitted; to calculate the weight of CO_2 emitted, multiply by 3.7.

50. Barbara Barrett, McClatchy Newspapers, "Little Doubt on Cause of Global Warming, Experts Tell Congress," February 8, 2007, available at http://www.mcclatchydc.com/staff/barbara_barrett/story/15581.html.
51. See note 1 in this chapter.
52. See note 38 in this chapter.
53. Intergovernmental Panel on Climate Change, *Climate Change 2007: Mitigation. Summary for Policymakers*, available at http://www.ipcc.ch/SPM040507.pdf (accessed September 16, 2007).
54. Intergovernmental Panel on Climate Change, *Climate Change 2007: Synthesis Report. Summary for Policymakers*, available at http://www.ipcc.ch/pdf/assessment-report/ar4/syr/ar4_syr_spm.pdf (accessed February 11, 2008).
55. T.M.L. Wigley, "A Combined Mitigation/Geoengineering Approach to Climate Stabilization," *Science* 314 (2006): 452–454.
56. Wallace S. Broecker, "CO_2 Arithmetic," *Science* 315 (2007): 1371.
57. Competitive Enterprise Institute, "We Call It *Life*," available at http://www.cei.org/pages/co2.cfm (accessed September 16, 2007).
58. James Heath et al., "Rising Atmospheric CO_2 Reduces Sequestration of Root-Derived Soil Carbon," *Science* 309 (2005): 1711–1713. Anthony W. King, "Plant Respiration in a Warmer World," *Science* 312 (2006): 536–537.
59. Stephen P. Long et al., "Food for Thought: Lower-Than-Expected Crop Yield Stimulation with Rising CO_2 Concentrations," *Science* 312 (2006): 1918–1921.
60. Christian Körner et al., "Carbon Flux and Growth in Mature Deciduous Forest Trees Exposed to Elevated CO_2," *Science* 309 (2005): 1360–1362.
61. Christopher B. Field, Michael J. Behrenfeld, James T. Randerson, and Paul Falkowski, "Primary Production of the Biosphere: Integrating Terrestrial and Oceanic Components," *Science* 281 (1998): 237–240.
62. Govindasamy Bala et al., "Combined Climate and Carbon-Cycle Effects of Large-Scale Deforestation," *Proceedings of the National Academy of Sciences*, published online before print, April 17, 2007; available at http://www.pnas.org/cgi/content/abstract/0608998104v1 (accessed September 16, 2007). Ken Caldeira, "Why Being Green Raises the Heat," *New York Times*, January 16, 2007.
63. Dennis Normile, "Getting at the Roots of Killer Dust Storms," *Science* 317 (2007): 314–316.
64. David Beerling, *The Emerald Planet: How Plants Changed Earth's History* (Oxford: Oxford University Press, 2007).
65. Global Footprint Network, "National Footprints," available at http://www.footprintnetwork.org/gfn_sub.php?content=national_footprints (accessed September 16, 2007).
66. Energy Information Administration, "Environment: Energy-Related Emissions Data and Environmental Analyses," available at http://www.eia.doe.gov/environment.html (accessed September 16, 2007). Total emissions: http://www.eia.doe.gov/pub/international/iealf/tableh1co2.xls. Per capita emissions: http://www.eia.doe.gov/pub/international/iealf/tableh1cco2.xls.
67. Richard Coniff, "Counting Carbons," *Discover*, August 2005, 54–61.

Chapter 4. Shade

1. M. Ortuäno, J. Alarcâon, E. Nicolâas, and A. Torrecillas, "Water Status Indicators of Lemon Trees in Response to Flooding and Recovery," *Biologia Plantarum* 51 (2007): 292–296.

2. Stanley A. Rice and John McArthur, "Water Flow through Xylem: An Investigation of a Fluid Dynamics Principle Applied to Plants," *American Biology Teacher* 66 (2004): 203–210.

3. For an example of measurements of cavitation during water stress, and whether or not recovery is possible, see M. J. Clearwater and C. J. Clark, "In Vivo Magnetic Resonance Imaging of Xylem Vessel Contents in Woody Lianas," *Plant, Cell and Environment* 26 (2003): 1205–1214.

4. J.M.O. Scurlock, G. P. Asner, and S. T. Gower, *Worldwide Historical Estimates of Leaf Area Index, 1932–2000*, December 2001 technical report to the Environmental Services Division, Oak Ridge National Laboratory, available at http://www.ornl.gov/~web-works/cppr/y2002/rpt/112600.pdf (accessed September 19, 2007).

5. D. A. Sampson, J. M. Vose, and H. L. Allen, "A Conceptual Approach to Stand Management Using Leaf Area Index as the Integral of Site Structure, Physiological Function, and Resource Supply," in *Ninth Biennial Southern Silvicultural Research. Conference*, USDA Forest Service, Southern Research Station General Technical Report SRS-20, 447–451; available at http://cwt33.ecology.uga.edu/publications/1233.pdf (accessed September 19, 2007).

6. Martin H. Zimmerman, *Xylem Structure and the Ascent of Sap* (Berlin: Springer-Verlag, 1983).

7. Christopher B. Field, Michael J. Behrenfeld, James T. Randerson, and Paul Falkowski, "Primary Production of the Biosphere: Integrating Terrestrial and Oceanic Components," *Science* 281 (1998): 237–240.

8. Nancy K. Wieland and Fakhri A. Bazzaz, "Physiological Ecology of Three Codominant Successional Annuals," *Ecology* 56 (1975): 681–688.

9. Goddard Space Flight Center, Scientific Visualization Studio, "Urban Growth Seen from Space," available at http://svs.gsfc.nasa.gov/stories/AAAS (accessed September 19, 2007). Laura Naranjo, "Seeing the City for the Trees," Earth Science Data and Services, National Aeronotics and Space Administration, available at http://nasadaacs.eos.nasa.gov/articles/2004/2004_urban.html (accessed September 17, 2007).

10. Goddard Space Flight Center, National Aeronautics and Space Administration, "NASA Satellite Confirms Urban Heat Islands Increase Rainfall around Cities," available at http://www.gsfc.nasa.gov/goddardnews/20020621/urbanrain.html (accessed September 19, 2007). Rob Gutro, "There's a Change in Rain around Desert Cities," Earth Observatory, National Aeronautics and Space Administration, available at http://earth-observatory.nasa.gov//Newsroom/NasaNews/2006/2006072522726.html (accessed September 19, 2007).

11. Virginia Gorsevski, Haider Taha, Dale Quattrochi, and Jeff Luval, "Air Pollution Prevention through Urban Heat Island Mitigation: An Update on the Urban Heat Island Pilot Project," available at http://www.ghcc.msfc.nasa.gov/uhipp/epa_doc.pdf (accessed September 19, 2007).

12. Environmental Energy Technologies, Lawrence Berkeley Laboratory, "Vegetation," available at http://eetd.lbl.gov/HeatIsland/Vegetation (accessed September 19, 2007).

13. U.S. Green Building Council, "Buildings and Climate Change," available at http://www.usgbc.org/DisplayPage.aspx?CMSPageID=1617 (accessed September 19, 2007). Architects at Technion-Israel Institute of Technology have designed an apartment building that uses recycled water and rainwater to irrigate a 100–square foot garden that comes with each apartment; see American Technion Society, available at http://www.ats.org/agro (accessed February 21, 2008).

14. David K. Randall, "Perhaps Only God Can Make a Tree, but Only People Could Put a Dollar Value on It," *New York Times*, April 18, 2007.
15. David B. Haines, "The Value of Trees in City of New Berlin," available at http://gis.esri.com/library/userconf/proc01/professional/papers/pap1021/p1021.htm (accessed September 19, 2007).
16. Nigel Dunnett and Noel Kingsbury, *Planting Green Roofs and Living Walls* (Portland, Ore.: Timber Press, 2004). Mark Gaulin, "The Green Building Advantage—Green Roofs Are the Way to Grow," available at http://www.greenroof.com/image_body/pdf/17–18_Reprint_02.pdf (accessed September 19, 2007). Mark Fischetti, "Living Cover," *Scientific American*, May 2008, 104–105.
17. Greenroofs.com, "Greenroof Projects Database," available at http://www.greenroofs.com/projects/plist.php (accessed September 19, 2007).
18. Sonnet L'Abbé, "Green Roofs in Winter: Hot Design for a Cold Climate," available at http://www.news.utoronto.ca/bin6/051117-1822.asp (accessed September 19, 2007).
19. A. Faber Taylor, Frances E. Kuo, and W. C. Sullivan, "Coping with ADD: The Surprising Connection to Green Play Settings," *Environment and Behavior* 33 (2001): 54–77. Charles A. Lewis, *Green Nature, Human Nature: The Meaning of Plants in Our Lives* (Urbana: University of Illinois Press, 1996).
20. Edward O. Wilson, *The Future of Life* (New York: Knopf, 2002), 139–140.

Chapter 5. The Water Cycle

1. As presented in Charles L. Redman, *Human Impact on Ancient Environments* (Tucson: University of Arizona Press, 1999), 101.
2. United Nations Food and Agricultural Organization, and Center for International Forestry Research, "Forests and Floods: Drowning in Fiction or Thriving on Facts?" available at ftp://ftp.fao.org/docrep/fao/008/ae929e/ae929e00.pdf (accessed September 20, 2007).
3. J. M. Bosch and J. D. Hewlett, "A Review of Catchment Experiments to Determine the Effect of Vegetation Changes on Water Yield and Evapotranspiration," *Journal of Hydrology* 103 (1982): 232–333. L. A. Bruijnzeel, "Predicting the Hydrological Impacts of Land Cover Transformation in the Humid Tropics: The Need for Integrated Research," in J.H.C. Gash et al., eds., *Amazonian Deforestation and Climate* (New York: Wiley, 1996).
4. Gene E. Likens et al., "Effects of Forest Cutting and Herbicide Treatment on Nutrient Budgets in the Hubbard Brook Watershed-Ecosystem," *Ecological Monographs* 40 (1970): 23–47; data available at http://tiee.ecoed.net/vol/v1/data_sets/hubbard/hubbard_faculty.xls (accessed September 21, 2007).
5. Tim Burt and Wayne Swank, "Forests or Floods?" *Geography Review*, May 2002, 37–41.
6. Marcos Heil Costa, Aurélie Botta, and Jeffrey A. Cardille, "Effects of Large-Scale Changes in Land Cover on the Discharge of the Tocantins River, Southeastern Amazonia," *Journal of Hydrology* 283 (2003): 206–217; available at http://www.sage.wisc.edu/pubs/articles/A-E/Costa/Costa2003JHydro.pdf (accessed September 20, 2007).
7. Ellen M. Douglas et al., "Policy Implications of a Pan-Tropic Assessment of the Simultaneous Hydrological and Biodiversity Impacts of Deforestation," in Eric Craswell

et al., eds., *Integrated Assessment of Water Resources and Global Change: A North-South Analysis* (Dordrecht: SpringerNetherlands, 2007), 211–232.

8. L. A. Bruijnzeel, "Hydrological Functions of Tropical Forests: Not Seeing the Soil for the Trees?" *Agriculture, Ecosystems, and Environment* 104 (2004): 185–228; available at http://www.asb.cgiar.org/pdfwebdocs/Hydrological_functions.pdf (accessed September 20, 2007).

9. Roy C. Sidle et al., "Erosion Processes in Steep Terrain—Truths, Myths, and Uncertainties Related to Forest Management in Southeast Asia," *Forest Ecology and Management* 224 (2006): 199–225.

10. S. Miura et al., "Protective Effect of Floor Cover against Soil Erosion on Steep Slopes Forested with *Chamaecyparis obtusa* (Hinoki) and Other Species," *Journal of Forest Research* 8 (2003): 27–35.

11. Unna Chokkalingam et al., eds., *Learning Lessons from China's Forest Rehabilitation Efforts: National Level Review and Special Focus on Guangdong Province*, Center for International Forestry Research, 2006; available at http://www.cifor.cgiar.org/publications/pdf_files/Books/Bchokkalingam0603.pdf (accessed September 20, 2007).

12. Plato, *Timaeus and Critias* (New York: Penguin, 1972).

13. Yavinder Malhi et al., "Climate Change, Deforestation, and the Fate of the Amazon," *Science* 319 (2008): 169–172.

14. Taikan Oki and Shinjiro Kanae, "Global Hydrological Cycles and World Water Resources," *Science* 313 (2006): 1068–1072.

15. John Perlin, *A Forest Journey: The Role of Wood in the Development of Civilization* (Cambridge, Mass.: Harvard University Press, 1989), 65.

16. Ferdinand Columbus, *The Life of the Admiral Christopher Columbus, by His Son*, trans. Benjamin Keen (New Brunswick, N.J.: Rutgers University Press, 1992), 143. *The Annual of Scientific Discovery, or, Year-Book of Facts in Science and Art*, published in 1852, contained a long entry on the relationship between vegetation cover and humidity, springs, and rivers.

17. Jared Diamond, *Collapse: How Societies Choose to Fail or Succeed* (New York: Viking, 2005), chap. 2.

18. Patrick V. Kirch, "Oceanic Islands: Microcosms of Global Change," chap. 1 in Charles L. Redman, Steven R. James, Paul R. Fish, and J. Daniel Rogers, eds., *The Archaeology of Global Change: The Impact of Humans on Their Environment* (Washington, D.C.: Smithsonian Institution Press, 2004).

19. Terry L. Hunt, "Rethinking the Fall of Easter Island," *American Scientist* 94 (2006). See also Jared Diamond, "Easter Island Revisited," *Science* 317 (2007): 1692–1694.

20. P. G. Bahn and J. R. Flenley, *Easter Island, Earth Island* (New York: Thames and Hudson, 1991).

21. Sonal Noticewala, "At Australia's Bunny Fence, Variable Cloudiness Prompts Climate Study," *New York Times*, August 14, 2007.

22. J. C. Shukla, C. Nobre, and P. Sellers, "Amazon Deforestation and Climate Change," *Science* 247 (1990): 1322–1325. G. K. Walker, Y. C. Sud, and R. Atlas, "Impact of the Ongoing Amazonian Deforestation on Local Precipitation: A GCM Simulation Study," *Bulletin of the American Meteorological Society* 76 (1995): 346–361.

23. U. S. Nair et al., "Impact of Land Use on Costa Rican Tropical Montane Cloud Forests: Sensitivity of Cumulus Cloud Field Characteristics to Lowland Deforestation," *Journal of Geophysical Research* 108 (2003): 4206.

24. N. Zeng, "Drought in the Sahel," *Science* 302 (2003): 999–1000.

25. Line J. Gordon et al., "Human Modification of Global Water Vapor Flows from the Land Surface," *Proceedings of the National Academy of Sciences* 102 (2005): 7612–7617. A. Kleidon, K. Fraedrich, and M. Heimann, "A Green Planet versus a Desert World: Estimating the Maximum Effect of Vegetation on the Land Surface Climate," *Climate Change* 44 (2000): 417–493.

26. Data available from the United Nations Food and Agriculture Organization at http://www.fao.org/nr/water/aquastat/regions/Africa/index.stm and http://www.fao.org/docrep/W4356E/w4356e06.HTM.

27. Gerrit Schoups et al., "Sustainability of Irrigated Agriculture in the San Joaquin Valley, California," *Proceedings of the National Academy of Sciences USA* 102 (2005): 15352–15356. Line J. Gordon et al., "Land Cover Change and Water Vapour Flows: Learning from Australia," *Philosophical Transactions of the Royal Society of London B (Biology)* 358 (2003): 1973–1984.

28. Rob Ferguson, *The Devil and the Disappearing Sea: A True Story about the Aral Sea Catastrophe* (Vancouver: Raincoast Books, 2005). Philip Micklin and Nikolay V. Aladin, "Reclaiming the Aral Sea," *Scientific American*, April 2008, 64–71.

29. Thorkild Jacobson and Robert M. Adams, "Salt and Silt in Ancient Mesopotamian Agriculture: Progressive Changes in Soil Salinity and Sedimentation Contributed to the Breakup of Past Civilizations," *Science* 128 (1958): 1251–1258.

30. Patricia L. Fall et al., "Environmental Impacts of the Rise of Civilization in the Southern Levant," chap. 7 in Redman, *Archaeology of Global Change*.

31. H. E. Dregne and N-T. Chou, "Global Desertification Dimensions and Costs," in H. E. Dregne, ed., *Degradation and Restoration of Arid Lands* (Lubbock: Texas Tech University Press, 1992).

32. Sandra Postel, *Pillar of Sand: Can the Irrigation Miracle Last?* (New York: Norton, 1999).

Chapter 6. Plants Feed the World

Epigraph: Thomas Jefferson, letter to Charles Willson Peale, August 20, 1811, in J. Jefferson Looney, ed., *Papers of Thomas Jefferson, Retirement Series* (Princeton, N.J.: Princeton University Press, 2007), 4:93.

1. Eric Schlosser, *Fast Food Nation* (New York: Harper Perennial, 2005).

2. Morgan Spurlock, *Don't Eat This Book: Fast Food and the Supersizing of America* (New York: Penguin, 2006).

3. Michael Pollan, *The Omnivore's Dilemma: A Natural History of Four Meals* (New York: Penguin, 2006). Natalie Angier, *The Canon: A Whirligig Tour of the Beautiful Basics of Science* (New York: Houghton Mifflin, 2007).

4. Ray L. Lindeman, "The Trophic-Dynamic Aspect of Ecology," *Ecology* 23 (1942): 399–418.

5. David Pimentel, "Livestock Production: Energy Inputs and the Environment," *Canadian Society of Animal Science Proceedings* 47 (1997): 17–26. David Pimentel and Marcia Pimentel, "Sustainability of Meat-Based and Plant-Based Diets and the Environment," *American Journal of Clinical Nutrition* 78 (2003): 660–663 (supplement). Available at http://www.ajcn.org/cgi/content/full/78/3/660S?ck=nck (accessed September 22, 2007).

6. See note 5. See also Brian Halweil and Danielle Nierenberg, "Meat and Seafood: The Global Diet's Most Costly Ingredients," chap. 5 in Worldwatch Institute, *State of the World: Innovations for a Sustainable Economy* (New York: Norton, 2008).

7. Pollan, *Omnivore's Dilemma*.
8. Bill McKibben, *Deep Economy: The Wealth of Communities and the Durable Future* (New York: Henry Holt, 2007).
9. Martin H. Bender, "Energy in Agriculture: Lessons from the Sunshine Farm Project," available at http://www.landinstitute.org/vnews/display.v/ART/2002/09/24/3dbeba6338ac3 (accessed September 22, 2007).
10. Jerry D. Glover, Cindy M. Cox, and John P. Reganold, "Future Farming: A Return to Roots?" *Scientific American*, August 2007, 82–89. Wes Jackson, "Natural Systems Agriculture: A Radical Alternative," *Agriculture, Ecosystems and Environment* 88 (2002): 111–117; available at http://www.landinstitute.org/vnews/display.v/ART/2001/04/17/3aa80bec9 (accessed September 22, 2007).
11. T. S. Cox et al., "Prospects for Developing Perennial Grains," *BioScience* 56 (2006): 649–659; available at http://www.landinstitute.org/pages/Bioscience_PerennialGrains.pdf (accessed September 22, 2007).
12. LocalHarvest, Inc., "LocalHarvest: Real Food, Real Farmers, Real Community," available at http://www.localharvest.org (accessed February 18, 2008).
13. Organic produce usually has higher levels of vitamin C than conventional produce, partly because the plants manufacture polyphenols that protect them from insects; crops on which pesticide is used, however, do not make as much of the defensive (and nutritious) compounds. See Danny K. Asami et al., "Comparison of Total Phenolic and Ascorbic Acid Content of Freeze-Dried and Air-Dried Marionberry, Strawberry, and Corn Using Conventional, Organic, and Sustainable Agricultural Practices," *Journal of Agricultural and Food Chemistry* 51 (2003): 1237–1241. Other nutritional differences between organic and conventional produce are less clear. See C. M. Williams, "Nutritional Quality of Organic Food: Shades of Grey or Shades of Green?" *Proceedings of the Nutrition Society* 61 (2002): 19–24. In addition, I personally am working with researchers at the U.S. Department of Agriculture to document the nutritional quality of organic produce in Oklahoma.
14. Michael Specter, "Big Foot," *New Yorker*, February 25, 2008; available at http://www.newyorker.com/reporting/2008/02/25/080225fa_fact_specter (accessed February 21, 2008).
15. Barbara Kingsolver, *Small Wonder* (New York: HarperCollins, 2002), 118.
16. T. E. Crews, "Perennial Crops and Endogenous Nutrient Supplies," *Renewable Agriculture and Food Systems* 20 (2005): 25–37.

Chapter 7. Plants Create Soil

Epigraph: Franklin Delano Roosevelt, "Letter to All State Governors on a Uniform Soil Conservation Law," February 26, 1937, United States Department of Agriculture, Natural Resources Conservation Service, "Soil Quotations," available at http://soils.usda.gov/education/resources/k_12/quotes.
1. Michael H. Glantz, ed., *Drought Follows the Plow* (New York: Cambridge University Press, 1994). The title implies that agriculture, through soil erosion and the elimination of the native plant cover, encourages droughts. The contributors claim that soil degradation increases vulnerability to drought rather than actually reducing rainfall.
2. United Nations Food and Agriculture Organization, "Terrastat," available at http://www.fao.org/ag/agl/agll/terrastat (accessed September 22, 2007). Data are not always reliable because they depend on government reporting for each country. Most observers

would not believe that Israel and Zimbabwe, for example, have 0 percent of their land degraded by erosion.

3. David R. Montgomery, *Dirt: The Erosion of Civilizations* (Berkeley: University of California Press, 2007). Daniel Hillel, *Out of the Earth: Civilization and the Life of the Soil* (Berkeley: University of California Press, 1992).

4. Jerry D. Glover, Cindy M. Cox, and John P. Reganold, "Future Farming: A Return to Roots?" *Scientific American*, August 2007, 82–89.

5. Wendell Berry, *The Unsettling of America: Culture and Agriculture* (Berkeley: Sierra Club Books/University of California Press, 1996). Latest of many reissues.

6. U.S. Department of Agriculture, National Agricultural Library, "Hyperaccumulators." A bibliography available at http://www.nal.usda.gov/wqic/Bibliographies/hypera.html (accessed September 22, 2007).

Chapter 8. Plants Create Habitats

1. David Beerling, *The Emerald Planet: How Plants Changed Earth's History* (Oxford: Oxford University Press, 2007).

2. David Berreby, "Running on Tundra," *Discover*, June 1996.

3. Peter Thomas, *Trees: Their Natural History* (Cambridge: Cambridge University Press, 2001).

4. Edward O. Wilson, *The Diversity of Life* (Cambridge, Mass.: Harvard University Press, 1992). Margaret Lowman, *Life in the Treetops: Adventures of a Woman in Field Biology* (New Haven, Conn.: Yale University Press, 2000).

5. David L. Peterson, Edward G. Schreiner, and Nelsa M. Buckingham, "Gradients, Vegetation and Climate: Spatial and Temporal Dynamics in the Olympic Mountains, U.S.A.," *Global Ecology and Biogeography Letters* 6 (1997): 7–17.

6. R. Alan Black and Richard N. Mack, "*Tsuga canadensis* in Ohio: Synecological and Phytogeographical Relationships," *Plant Ecology* 32 (1976): 11–19.

7. Doria R. Gordon and Kevin J. Rice, "Competitive Suppression of *Quercus douglasii* (Fagaceae) Seedling Emergence and Growth," *American Journal of Botany* 87 (2000): 986–994.

8. United States Forest Service, "Blue Oak," available at http://www.na.fs.fed.us/pubs/silvics_manual/volume_2/quercus/douglasii.htm (accessed September 23, 2007).

9. Louis R. Iverson and Anantha M. Prasad, "Potential Redistribution of Tree Species Habitats under Five Climate Change Scenarios in the Eastern U.S.," *Forest Ecology and Management* 155 (2002): 205–222. Louis R. Iverson and Anantha M. Prasad, "Predicting Abundance of Eighty Tree Species Following Climate Change in the Eastern United States," *Ecological Monographs* 68 (1998): 465–485.

10. Thomas, *Trees*.

11. Cornelia Dean, "The Preservation Predicament," *New York Times*, January 29, 2008.

12. Douglas Fox, "Back to the No-Analog Future?" *Science* 316 (2007): 823–825. Raymond E. Gullison et al., "Tropical Forests and Climate Policy," *Science* 316 (2007): 985–986.

Chapter 9. Plants Heal the Landscape

Epigraph: Theodore Roosevelt, Arbor Day Proclamation, 1907.

1. Phillip L. Sims, "Grasslands," chap. 9 in Michael G. Barbour and William Dwight Billings, eds., *North American Terrestrial Vegetation* (New York: Cambridge University Press, 1988).

2. Neil E. West, "Intermountain Deserts, Shrub Steppes, and Woodlands," chap. 7 in Barbour and Billings, *North American Terrestrial Vegetation*.

3. Michael G. Barbour et al., "Fire," chap. 16 in *Terrestrial Plant Ecology* (Menlo Park, Calif.: Addison Wesley/Benjamin Cummings, 1999).

4. Jon E. Keeley and Sterling C. Keeley, "Chaparral," chap. 6 in Barbour and Billings, *North American Terrestrial Vegetation*.

5. Jon E. Keeley, "Demographic Structure of California Chaparral in the Long-Term Absence of Fire," *Journal of Vegetation Science* 3 (1992): 79–90.

6. Jon E. Keeley and C. D. Fotheringam, "Smoke-Induced Seed Germination in Californian Chaparral," *Science* 276 (1997): 1248–1250.

7. P. Grogan et al., "Below-Ground Ectomycorrhizal Community Structure in a Recently Burned Bishop Pine Forest," *Journal of Ecology* 88 (2000): 1051–1062.

8. Henry Chandler Cowles, "The Ecological Relations of the Vegetation on the Sand Dunes of Lake Michigan," *Botanical Gazette* 27 (1899): 95–117, 167–202, 281–308, 361–391.

9. John Lichter, "Colonization Constraints during Primary Succession on Coastal Lake Michigan Sand Dunes," *Journal of Ecology* 88 (2000): 825–839.

10. Virginia H. Dale, Frederick J. Swanson, and Charles M. Crisafulli, eds., *Ecological Responses to the 1980 Eruption of Mount St. Helens* (New York: Springer, 2005). Carol Kaesuk Yoon, "As Mt. St. Helens Recovers, Old Wisdom Crumbles," *New York Times*, May 16, 2000.

11. Fakhri A. Bazzaz, "Succession on Abandoned Fields in the Shawnee Hills, Southern Illinois," *Ecology* 49 (1968): 924–936. Steward T. A. Pickett, "Population Patterns through Twenty Years of Oldfield Succession," *Plant Ecology* 49 (1982): 45–59.

12. Kristine N. Hopfensperger, "A Review of Similarity between Seed Bank and Standing Vegetation across Ecosystems," *Oikos* 116 (2007): 1438–1448.

13. Fakhri A. Bazzaz, "The Physiological Ecology of Plant Succession," *Annual Review of Ecology and Systematics* 10 (1979): 351–371. Steward T. A. Pickett, "Succession: An Evolutionary Interpretation," *American Naturalist* 110 (1976): 107–119. Fakhri A. Bazzaz, *Plants in Changing Environments: Linking Physiological, Population, and Community Ecology* (New York: Cambridge University Press, 1996).

14. David S. Gill and Peter L. Marks, "Tree and Shrub Seedling Colonization of Old Fields in Central New York," *Ecological Monographs* 61 (1991): 183–205.

15. Christian Körner et al., "Small Differences in Arrival Time Influence Composition and Productivity of Plant Communities," *New Phytologist* 177 (2008): 698–705. Norman L. Christensen and Robert K. Peet, "Convergence during Secondary Forest Succession," *Journal of Ecology* 72 (1984): 25–36.

16. David A. Wardle et al., "The Response of Plant Diversity to Ecosystem Retrogression: Evidence from Contrasting Long-Term Chronosequences," *Oikos* 117 (2008): 93–103.

17. Jonathan Adams, "North America during the Last 150,000 Years," Oak Ridge National Laboratory, available at http://www.esd.ornl.gov/projects/qen/nercNORTHAMERICA .html (accessed February 16, 2008).

18. E. C. Pielou, *After the Ice Age: Return of Life to Glaciated North America* (Chicago: University of Chicago Press, 1992).

19. U.S. Forest Service, Rocky Mountain Research Station, "Restoration Ecology of Disturbed Lands," available at http://www.fs.fed.us/rm/logan/4301 (accessed February 18, 2008).

20. Fakhri A. Bazzaz and Steward T. A. Pickett, "Physiological Ecology of Tropical Succession: A Comparative Review," *Annual Review of Ecology and Systematics* 11 (1980): 287–310.

21. Sergio Guevara et al., "The Role of Remnant Forest Trees in Tropical Secondary Succession," *Plant Ecology* 66 (1986): 77–84. Karen D. Holl, "Effect of Shrubs on Tree Seedling Establishment in an Abandoned Tropical Pasture," *Journal of Ecology* 90 (2002): 179–187. Karen D. Holl et al., "Tropical Montane Forest Restoration in Costa Rica: Overcoming Barriers to Dispersal and Establishment," *Restoration Ecology* 8 (2000): 339–349.
22. Yavinder Malhi et al., "Climate Change, Deforestation, and the Fate of the Amazon," *Science* 319 (2008): 169–172.
23. U.S. Department of Agriculture, National Agricultural Library, "National Invasive Species Information Center," available at *http://www.invasivespeciesinfo.gov* (accessed September 23. 2007).

Chapter 10. How Agriculture Changed the World

Epigraph: Statement allegedly made by Jefferson at a dinner. See the Monticello Web site at http://www.monticello.org/gardens/grounds/trees.html (accessed September 25, 2007).
1. Michael Pollan, *The Botany of Desire: A Plant's-Eye View of the World* (New York: Random House, 2001). Michael Pollan, *The Omnivore's Dilemma: A Natural History of Four Meals* (New York: Penguin, 2006).
2. Michael Balter, "Seeking Agriculture's Ancient Roots," *Science* 316 (2007): 1830–1835. Ken-Ichi Tanno and George Willcox, "How Fast Was Wild Wheat Domesticated?" *Science* 311 (2006): 1886. Ehud Weiss, Mordechai E. Kislev, and Anat Hartmann, "Autonomous Cultivation before Domestication," *Science* 312 (2006): 1608–1610. David Rindos, *The Origins of Agriculture: An Evolutionary Perspective* (San Diego: Academic Press, 1984).
3. Jared Diamond, *Guns, Germs, and Steel: The Fates of Human Societies* (New York: Norton, 1997).
4. It appears that the major transformation of the growth form of teosinte (*Zea mays* ssp. *parviglumis*) into the growth form of maize (*Zea mays* ssp. *mays*) occurred because of a mutation in a single gene. See J. Doebley, A. Stec, and L. Hubbard, "The Evolution of Apical Dominance in Maize," *Nature* 386 (1997): 485–488. However, many centuries of artificial selection were required to transform primitive maize, with tiny ears of grain, into the large-eared varieties that the New World natives possessed at the time of Columbus.
5. Henry Hobhouse, *Seeds of Change: Five Plants That Transformed Mankind* (New York: Harper and Row, 1987). Hobhouse describes other examples as well, such as quinine, cotton, and tea.
6. Sidney W. Mintz, *Sweetness and Power: The Place of Sugar in Modern History* (New York: Penguin, 1986).
7. Mary Kilbourne Matossian, *Poisons of the Past: Molds, Epidemics, and History* (New Haven, Conn.: Yale University Press, 1991). Redcliffe N. Salaman, *The History and Social Influence of the Potato* (Cambridge: Cambridge University Press, 1949).
8. Jack Weatherford, *Indian Givers: How the Indians of the Americas Transformed the World* (New York: Ballantine, 1989).
9. Nelson Foster and Linda S. Cordell, eds., *Chilies to Chocolate: Food the Americas Gave the World* (Tucson: University of Arizona Press, 1992).
10. The Old Testament contains a recipe for bread that contains both grain and legume flours (Ezekiel 4:9). The prophet Ezekiel supposedly survived for 390 days while eating nothing but this bread.

11. Agroforestry Net, Inc., "Agroforestry.net," available at http://www.agroforestry.net (accessed February 21, 2008). U.S. Department of Agriculture, "National Agroforestry Center," available at http://www.unl.edu/nac/ (accessed February 21, 2008). World Agroforestry Centre, "Agroforestry Addresses Climate Change and Poverty," available at http://www.worldagroforestry.org (accessed February 21, 2008). National Sustainable Agriculture Information Service, "Agroforestry Overview," available at http://www.attra.org/attra-pub/agroforestry.html (accessed February 21, 2008).

Chapter 11. Why We Need Plant Diversity

Epigraph: Rachel Carson, *Silent Spring* (1962; repr., New York: Houghton Mifflin, 2002).
1. "Hansen Files Landmark Bill to Restore Original Intent of Endangered Species Act," *Sierra Times*, November 13, 2002, available at http://www.sierratimes.com/02/11/14/arst111402.htm (accessed September 23, 2007).
2. Mark Plotkin, *Tales of a Shaman's Apprentice: An Ethnobotanist Searches for New Medicines in the Amazon Rain Forest* (New York: Viking, 1993). Nicole Maxwell, *Witch-Doctor's Apprentice: Hunting for Medicinal Plants in the Amazon* (New York: Citadel, 1961).
3. United Nations Food and Agriculture Organization, "FAOSTAT," available at http://faostat.fao.org/site/340/default.aspx (accessed September 23, 2007).
4. Gary Paul Nabhan, *Why Some Like It Hot: Food, Genes, and Cultural Diversity* (Washington, D.C.: Island Press, 2004).
5. Barbara Kingsolver, *Small Wonder* (New York: HarperCollins, 2002), 100.
6. Daniel Charles, "A 'Forever' Seed Bank Takes Root in the Arctic," *Science* 312 (2006): 1730–1731. Elizabeth Rosenthal, "Near Arctic, Seed Vault Is a Fort Knox of Food," *New York Times*, February 29, 2008.
7. Seed Savers Exchange, "Seed Savers Exchange," available at http://www.seedsavers.org (accessed September 23, 2007). Southern Exposure Seed Exchange, "Southern Exposure Seed Exchange," available at http://www.southernexposure.com (accessed September 23, 2007).
8. Nigel J. H. Smith et al., *Tropical Forests and Their Crops* (Ithaca, N.Y.: Cornell University Press, 1992).
9. George Ray McEachern, "A Texas Grape and Wine History," available at http://aggie-horticulture.tamu.edu/southerngarden/Texaswine.html (accessed September 23, 2007).
10. Smith, *Tropical Forests and Their Crops*.
11. U.S. Department of the Army. *U.S. Army Survival Manual FM 21-76*, 1998. Distributed by Dorset Press, New York.
12. Thomas Elias and Peter Dykeman, *Edible Wild Plants: A North American Field Guide* (New York: Sterling, 1990).
13. C. Wiley Hinman and Jack W. Hinman, *The Plight and Promise of Arid Land Agriculture* (New York: Columbia University Press, 1992).
14. City of Arcata Environmental Services, "Arcata Marsh and Wildlife Sanctuary," available at http://www.cityofarcata.org/index.php?option=com_content&task=view&id=20&Itemid=47 (accessed September 23, 2007).
15. Jerry D. Glover, Cindy M. Cox, and John P. Reganold, "Future Farming: A Return to Roots?" *Scientific American*, August 2007, 82–89. T. S. Cox et al., "Breeding Perennial Grain Crops," *Critical Reviews in Plant Sciences* 21 (2002): 59–91.
16. National Research Council, Board on Science and Technology for International Development, *Lost Crops of Africa. Volume 1: Grains* (Washington, D.C.: National Academy Press, 1996).

17. Dan Koeppel, *Banana: The Fate of the Fruit That Changed the World* (New York: Hudson Street Press, 2007). Steve Striffler and Mark Moberg, eds., *Banana Wars: Power, Production, and History in the Americas* (Durham, N.C.: Duke University Press, 2003). Peter Chapman, *Bananas! How the United Fruit Company Shaped the World* (Edinburgh: Canongate, 2008).

18. Wes Jackson, *New Roots for Agriculture* (Lincoln: University of Nebraska Press, 1980).

19. Smith, *Tropical Forests and Their Crops.*

Chapter 12. What Can We Do?

Epigraph: Upton Sinclair, *I, Candidate for Governor: And How I Got Licked* (1935; repr., Berkeley: University of California Press, 1994).

1. Wangari Maathai, National Public Radio interview, September 27, 2007. See also Wangari Maathai, *Unbowed: A Memoir* (New York: Anchor, 2007). See also Green Belt Movement, "The Green Belt Movement," available at http://www.greenbeltmovement.org (accessed October 2, 2007). See also United Nations Environmental Programme, "The Billion Tree Campaign," available at http://www.unep.org/billiontreecampaign (accessed October 2, 2007).

2. Feng An and Amanda Sauer, *Comparison of Passenger Vehicle Fuel Economy and GHG Emission Standards around the World*, Pew Center on Global Climate Change, 2004, available at http://www.pewclimate.org/docUploads/Fuel%20Economy%20and%20GHG%20Standards_010605_110719.pdf (accessed October 1, 2007).

3. U.S. Department of Energy, Energy Efficiency and Renewable Energy, "Fueleconomy.gov," available at http://www.fueleconomy.gov (accessed October 1, 2007). See also Tom Clynes, "The Energy Fix: Ten Steps to End America's Fossil-Fuel Addiction," *Popular Science*, July 2006, 47–61.

4. Rocky Mountain Institute, "Profitable Innovations for Energy and Resource Efficiency," available at http://www.rmi.org (accessed October 1, 2007).

5. Colleen Long, "System Relies on Ice to Chill Buildings," Environmental News Network, July 16, 2007, available at http://www.enn.com.

6. Mark Jacobson and Gilbert M. Masters, "Exploiting Wind versus Coal," *Science* 293 (2001): 1438.

7. Renton Righelato and Dominick V. Spracklen, "Carbon Mitigation by Biofuels or by Saving and Restoring Forests?" *Science* 317 (2007): 902.

8. Joseph Fargione, et al., "Land Clearing and the Biofuel Carbon Debt," *Science* 319 (2008): 1235–1238. Timothy Searchinger, et al., "Use of U.S. Croplands for Biofuels Increases Greenhouse Gases through Emissions from Land Use Change," *Science* 319 (2008): 1238–1240.

9. David Tilman, Jason Hill, and Clarence Lehman, "Carbon-Negative Biofuels from Low-Input High-Diversity Grassland Biomass," *Science* 314 (2006): 1598–1600. Robert F. Service, "Biofuel Researchers Prepare to Reap a New Harvest," *Science* 315 (2007): 1488–1491. Matthew L. Wald and Alexei Barrionuevo, "Chasing a Dream Made of Weeds: As Corn Prices Soar, Pressure Builds to Make a Cheap Cellulosic Ethanol," *New York Times*, April 17, 2007.

10. Plug In America, "Plug In America," available at http://www.pluginamerica.com (accessed October 1, 2007).

11. Andrew Ross Sorkin, "A Buyout Deal That Has Many Shades of Green," *New York Times*, February 26, 2007.

12. H. Josef Hebert, "White House Official Singled Out for Editing Climate Reports to Work for ExxonMobil," Environmental News Network, June 15, 2005, available at http://www.enn.com.

13. Andrew C. Revkin, "A Young Bush Appointee Resigns His Post at NASA," *New York Times*, February 8, 2006.

14. Elizabeth Williamson, "Interior Department Official Facing Scrutiny Resigns," *Washington Post*, May 2, 2007.

15. H. Josef Hebert, "EPA Scientists Complain about Political Pressure," Environmental News Network, April 23, 2008, available at http://www.enn.com.

16. Edmund L. Andrews, "Inspector Finds Broad Failures in Oil Program," *New York Times*, September 26, 2007.

17. Eli Kintisch, "Industry Conservation Programs Face White House Cuts," *Science* 312 (2006): 675.

18. Chris Mooney, *The Republican War on Science* (New York: Basic Books, 2005). Carl Pope and Paul Rauber, *Strategic Ignorance: Why the Bush Administration Is Recklessly Destroying a Century of Environmental Progress* (Berkeley: University of California Press, 2004).

19. Eli Kintisch, "U.S. Supreme Court Gets Arguments for EPA to Regulate CO_2," *Science* 313 (2006): 1375.

20. City of Los Angeles, "Green LA: An Action Plan to Lead the Nation in Fighting Global Warming," May 2007, available at http://www.lacity.org/ead/EADWeb-AQD/GreenLA_CAP_2007.pdf (accessed October 2, 2007).

21. United States Climate Action Partnership, "USCAP: United States Climate Action Partnership," available at http://www.us-cap.org (accessed February 22, 2008). Timothy Gardner, "Big Investors Urge U.S. to Slash CO_2 Emissions," Environmental News Network, March 20, 2007, available at http://www.enn.com.

22. Thomas Wagner, "Business World Finally Sees Potential Profits in Joining the Battle against Global Warming," Environmental News Network, August 17, 2007, available at http://www.enn.com.

23. Sharon Begley, "The Truth about Denial," *Newsweek*, August 13, 2007, 20–29.

24. Christiansandclimate.org, "Evangelical Climate Initiative," available at http://www.christiandandclimate.org (accessed October 1, 2007). Eric Gorski, "Southern Baptists Question Human Role in Global Warming; Say Regulations Hurt Poor," Environmental News Network, June 14, 2007, available at http://www.enn.com.

25. Steve Hargreaves, "Exxon Linked to Climate Change Pay Out: Think Tank Offers Scientists $10,000 to Criticize UN Study Confirming Global Warming and Placing Blame on Humans," *CNN Money*, February 5, 2007; available at http://money.cnn.com/2007/02/02/news/companies/exxon_science/index.htm (accessed October 1, 2007). As of October 2007, no information was available on the Web site of the American Enterprise Institute (http://www.aei.org).

26. John R. Christy, "The Global Warming Crisis," chap. 1 in Ronald Bailey, ed., *Global Warming and Other Eco-Myths: How the Environmental Movement Uses False Science to Scare Us to Death* (New York: Crown, 2002); see figure 1.1.

27. Elizabeth Royte, "The Gospel According to John," *Discover*, February 2001, available at http://discovermagazine.com/2001/feb/featgospel (accessed February 16, 2008).

28. Heartland Institute, "Heartland Institute," available at http://www.heartland.org. The list of scientists is at http://www.oism.org/pproject. The advertisement was in *New York Times*, February 19, 2008.

29. H. Josef Hebert, "House Approves Fees, Taxes on Oil Companies; Plans to Use Money for Renewable Fuels," Environmental News Network, January 19, 2007, available at http://www.enn.com.

30. Dave Gram, "Judge Rejects Carmakers' Emission Suit," Environmental News Network, September 12, 2007, available at http://www.enn.com.

31. Intergovernmental Panel on Climate Change, *Climate Change 2007: Synthesis Report. Summary for Policymakers*, available at http://www.ipcc.ch/pdf/assessment-report/ar4/syr/ar4_syr_spm.pdf (accessed February 16, 2008). Richard A. Kerr, "Global Warming: How Urgent Is Climate Change?" *Science* 318 (2007): 1230–1231.

32. Jeremy Lovell, "Britain Proposes Legal Limits on Carbon Emissions," Environmental News Network, March 14, 2007, available at http://www.enn.com.

33. Sara Kugler, "Sixteen Cities to Go Green under Clinton Plan," Environmental News Network, May 16, 2007, available at http://www.enn.com. See also William J. Clinton Foundation, "Clinton Climate Initiative," available at http://www.clintonfoundation.org/cf-pgm-cci-home.htm (accessed October 3, 2007).

34. Gerard Wynn, "World Bank Eyes $250 Million Deal to Save Forests," Environmental News Network, May 4, 2007, available at http://www.enn.com.

35. Energy Action Coalition, "Power Shift 2007," available at http://www.powershift07.org (accessed October 2, 2007).

36. John M. Broder and Marjorie Connelly, "Public Says Warming Is a Problem, but Remains Split on Response," *New York Times*, April 27, 2007.

37. Elisabeth Rosenthal, "Vatican Tree Penance: Forgive Us Our CO_2," *New York Times*, September 17, 2007.

38. Kevin Smith, "The Carbon Neutral Myth: Offset Indulgences for Your Climate Sins," Transnational Institute, Carbon Trade Watch, available at http://www.carbontrade-watch.org/pubs/carbon_neutral_myth.pdf (accessed October 2, 2007).

39. Dietrich Bonhoeffer, *The Cost of Discipleship* (New York: Macmillan, 1963). Peter Schweitzer, "Offset Away Our Guilt," *USA Today*, September 25, 2007. Michael Hill, "Do Trees Make It O.K. to Drive an SUV?" Environmental News Network, May 28, 2007, available at http://www.enn.com.

40. Norman Geisler, *Knowing the Truth about Creation: How It Happened and What It Means for Us* (Ann Arbor, Mich.: Servant Books, 1989).

41. Kevin Phillips, *American Theocracy: The Peril and Politics of Radical Religion, Oil, and Borrowed Money in the 21st Century* (New York: Penguin, 2007).

42. Bill McKibben, "It's Easy Being Green," *Mother Jones*, July/August 2002.

43. Clapp quoted in John Carey, "Taming the Oil Beast," *Business Week*, February 24, 2003, 98. Russia also uses its export of fossil fuels as a political force against neighboring nations.

44. Wangari Maathai, National Public Radio interview, September 27, 2007. See also Daniel Wallis, "Global Warming to Hit Poor the Worst, Says U.N.'s Ban," Environmental News Network, February 6, 2007, available at http://www.enn.com.

45. Brown quoted in Deborah Zabarenko, "U.N. Climate Change Meeting Aims at Rich Countries," Environmental News Network, July 31, 2007, available at http://www.enn.com.

46. Francis Hopkinson, *The Miscellaneous Essays and Occasional Writings of Francis Hopkinson* (Philadelphia, n.d.), 256.

47. Stevenson Center for Community and Economic Development, "Adlai E. Stevenson II," available at http://www.stevensoncenter.org/stevenson.htm (accessed June 5, 2008).

bibliography

Abrams, Marc D. "Red Maple Paradox." *BioScience* 48 (1998): 1–18.

Adams, Jonathan. "North America during the Last 150,000 Years." Oak Ridge National Laboratory. Available at http://www.esd.ornl.gov/projects/qen/nercNORTHAMERICA.html.

Agroforestry Net, Inc. "Agroforestry.net." Available at http://www.agroforestry.net.

An, Feng, and Amanda Sauer. *Comparison of Passenger Vehicle Fuel Economy and GHG Emission Standards around the World.* Pew Center on Global Climate Change, 2004. Available at http://www.pewclimate.org/docUploads/Fuel%20Economy%20and%20GHG%20Standards_010605_110719.pdf.

Anderson, Edgar. *Plants, Man, and Life.* London: Andrew Melrose, 1954.

Andrews, Edmund L. "Inspector Finds Broad Failures in Oil Program." *New York Times,* September 26, 2007.

Angier, Natalie. *The Canon: A Whirligig Tour of the Beautiful Basics of Science.* New York: Houghton Mifflin, 2007.

Asami, Danny K., et al. "Comparison of Total Phenolic and Ascorbic Acid Content of Freeze-Dried and Air-Dried Marionberry, Strawberry, and Corn Using Conventional, Organic, and Sustainable Agricultural Practices." *Journal of Agricultural and Food Chemistry* 51 (2003): 1237–1241.

Bahn, P. G., and J. R. Flenley. *Easter Island, Earth Island.* New York: Thames and Hudson, 1991.

Baker, Herbert G. *Plants and Civilization.* Belmont, Calif.: Wadsworth, 1965.

Bala, Govindasamy, et al. "Combined Climate and Carbon-Cycle Effects of Large-Scale Deforestation." *Proceedings of the National Academy of Sciences,* April 17, 2007. Available at http://www.pnas.org/cgi/content/abstract/0608998104v1.

Balter, Michael. "Seeking Agriculture's Ancient Roots." *Science* 316 (2007): 1830–1835.

Barbour, Michael G., et al. "Fire." Chapter 16 in *Terrestrial Plant Ecology.* Menlo Park, Calif.: Addison Wesley/Benjamin Cummings, 1999.

———. "Major Vegetation Types of North America." Chapter 20 in *Terrestrial Plant Ecology.* Menlo Park, Calif.: Addison Wesley/Benjamin Cummings, 1999.

Barbour, Michael G., and William Dwight Billings, eds. *North American Terrestrial Vegetation.* New York: Cambridge University Press, 1988.

Barnett, Tim P., et al. "Human-Induced Changes in the Hydrology of the Western United States." *Science* 319 (2008): 1080–1083.

———. "Penetration of Human-Induced Warming into the World's Oceans." *Science* 309 (2005): 284–287.

Barnola, J.-M., et al. "Historical Carbon Dioxide Record from the Vostok Ice Core." In *Trends: A Compendium of Data on Global Change,* Carbon Dioxide Information Analysis Center, Oak Ridge National Laboratory, 2000. Available at http://cdiac.ornl.gov/trends/co2/vostok.htm.

Bibliography

Barnosky, Anthony D. "Assessing the Causes of Late Pleistocene Extinctions on the Continents." *Science* 306 (2004): 70–75.

Barrett, Barbara. McClatchy Newspapers. "Little Doubt on Cause of Global Warming, Experts Tell Congress." February 8, 2007. Available at http://www.mcclatchydc.com/staff/barbara_barrett/story/15581.html.

Bartram, William. *Travels through North and South Carolina, Georgia, East and West Florida, the Cherokee Country, the Extensive Territories of the Muscogulges or Creek Confederacy, and the Country of the Chactaws.* 1791. Reprint, Savannah, Ga.: Beehive Press, 1973.

Bazzaz, Fakhri A. "The Physiological Ecology of Plant Succession." *Annual Review of Ecology and Systematics* 10 (1979): 351–371.

———. *Plants in Changing Environments: Linking Physiological, Population, and Community Ecology.* New York: Cambridge University Press, 1996.

———. "Succession on Abandoned Fields in the Shawnee Hills, Southern Illinois." Ecology 49 (1968): 924–936.

Bazzaz, Fakhri A., and Steward T. A. Pickett. "Physiological Ecology of Tropical Succession: A Comparative Review." *Annual Review of Ecology and Systematics* 11 (1980): 287–310.

Bearhop, Stuart, et al. "Assortative Mating as a Mechanism for Rapid Evolution of a Migratory Divide." *Science* 310 (2005): 502–504.

Bechmann, Roland. *Trees and Man: The Forest in the Middle Ages.* Translated by Katharyn Dunham. New York: Paragon House, 1990.

Beerling, David. *The Emerald Planet: How Plants Changed Earth's History.* Oxford: Oxford University Press, 2007.

Begley, Sharon. "The Truth about Denial." *Newsweek*, August 13, 2007, 20–29.

Behre, Karl-Ernst. "The Role of Man in European Vegetation History." In B. Huntley and T. Webb III, eds., *Vegetation History*, 633–672. Dordrecht: Kluwer Scientific Publications, 1988.

Bell, Robin E. "The Unquiet Ice." *Scientific American*, February 2008, 60–67.

Bender, Martin H. "Energy in Agriculture: Lessons from the Sunshine Farm Project." Available at http://www.landinstitute.org/vnews/display.v/ART/2002/09/24/3dbeba6338ac3.

Benton, Michael J. *When Life Nearly Died: The Greatest Mass Extinction of All Time.* London: Thames and Hudson, 2003.

Berreby, David. "Running on Tundra." *Discover*, June 1996.

Berry, Wendell. *The Unsettling of America: Culture and Agriculture.* Berkeley: Sierra Club Books/University of California Press, 1996.

Black, R. Alan, and Richard N. Mack. "*Tsuga canadensis* in Ohio: Synecological and Phytogeographical Relationships." *Plant Ecology* 32 (1976): 11–19.

Bohannon, John. "Microbe May Push Photosynthesis into Deep Water." *Science* 308 (2005): 1855.

Bonhoeffer, Dietrich. *The Cost of Discipleship.* New York: Macmillan, 1963.

Bonnicksen, Thomas M. *America's Ancient Forests: From the Ice Age to the Age of Discovery.* New York: Wiley, 2000.

Bosch, J. M., and J. D. Hewlett. "A Review of Catchment Experiments to Determine the Effect of Vegetation Changes on Water Yield and Evapotranspiration." *Journal of Hydrology* 103 (1982): 232–333.

Braasch, Gary. *Earth under Fire: How Global Warming Is Changing the World.* Berkeley: University of California Press, 2007.

Broder, John M., and Marjorie Connelly. "Public Says Warming Is a Problem, but Remains Split on Response." *New York Times*, April 27, 2007.

Broecker, Wallace S. "CO2 Arithmetic." *Science* 315 (2007): 1371.

Broecker, Wallace S., and Robert Kunzig. Fixing Climate: *What Past Climate Changes Reveal about the Current Threat—And How to Counter It.* New York: Farrar, Straus, Giroux, 2008.

Brower, David, with Steve Chapple. *Let the Mountains Talk, Let the Rivers Run: A Call to Those Who Would Save the Earth.* New York: HarperCollins, 1995.

Bruijnzeel, L. A. "Hydrological Functions of Tropical Forests: Not Seeing the Soil for the Trees?" *Agriculture, Ecosystems, and Environment* 104 (2004): 185–228. Available at http://www.asb.cgiar.org/pdfwebdocs/Hydrological_functions.pdf.

———. "Predicting the Hydrological Impacts of Land Cover Transformation in the Humid Tropics: The Need for Integrated Research." In J.H.C. Gash et al., eds., *Amazonian Deforestation and Climate.* New York: Wiley, 1996.

Brumfiel, Geoff. "Academy Affirms Hockey-Stick Graph." Available at http://www.nature.com/news/2006/060626/full/4411032a.html.

Burt, Tim, and Wayne Swank. "Forests or Floods?" *Geography Review,* May 2002, 37–41.

Caldeira, Ken. "Why Being Green Raises the Heat." *New York Times,* January 16, 2007.

Carey, John. "Taming the Oil Beast." *Business Week,* February 24, 2003, 98.

Cazenave, Anny. "How Fast Are the Ice Sheets Melting?" *Science* 314 (2006): 1250–1252.

Center for the Study of Carbon Dioxide and Global Change. "CO$_2$ Science." Available at http://www/co2science.org.

Chapman, Peter. *Bananas! How the United Fruit Company Shaped the World.* Edinburgh: Canongate, 2008.

Charles, Daniel. "A 'Forever' Seed Bank Takes Root in the Arctic." *Science* 312 (2006): 1730–1731.

Charmantier, Anne, et al. "Adaptive Phenotypic Plasticity in Response to Climate Change in a Wild Bird Population." *Science* 320 (2008): 800–803.

Chen, J. L., C. R. Wilson, and B. D. Tapley. "Satellite Gravity Measurements Confirm Accelerated Melting of Greenland Ice Sheet." *Science* 313 (2006): 1958–1960.

Chokkalingam, Unna, et al., eds. *Learning Lessons from China's Forest Rehabilitation Efforts: National Level Review and Special Focus on Guangdong Province.* Center for International Forestry Research, 2006. Available at http://www.cifor.cgiar.org/publications/pdf_files/Books/Bchokkalingam0603.pdf.

Christensen, Norman L., and Robert K. Peet. "Convergence during Secondary Forest Succession." *Journal of Ecology* 72 (1984): 25–36.

Christiansandclimate.org. "Evangelical Climate Initiative." Available at http://www.christiandandclimate.org.

Christy, John R. "The Global Warming Fiasco." Chapter 1 in Ronald Bailey, ed., *Global Warming and Other Eco-Myths: How the Environmental Movement Uses False Science to Scare Us to Death.* New York: Crown, 2002.

City of Arcata Environmental Services. "Arcata Marsh and Wildlife Sanctuary." Available at http://www.cityofarcata.org/index.php?option=com_content&task=view&id=20&Itemid=47.

City of Los Angeles. Green LA: An Action Plan to Lead the Nation in Fighting Global Warming. May 2007. Available at http://www.lacity.org/ead/EADWeb-AQD/GreenLA_CAP_2007.pdf.

Clearwater, M. J., and C. J. Clark. "In Vivo Magnetic Resonance Imaging of Xylem Vessel Contents in Woody Lianas." *Plant, Cell and Environment* 26 (2003): 1205–1214.

Clynes, Tom. "The Energy Fix: Ten Steps to End America's Fossil-Fuel Addiction." *Popular Science,* July 2006, 47–61.

Bibliography

CNA Corporation. "National Security and the Threat of Climate Change." Available at http://securityandclimate.cna.org/report.

Coe, Sophie D., and Michael D. Coe. *The True History of Chocolate*. London: Thames and Hudson, 1996.

Collins, William, et al. "The Physical Science behind Climate Change." *Scientific American*, August 2007, 64–73.

Columbus, Ferdinand. *The Life of the Admiral Christopher Columbus, by His Son*. Translated by Benjamin Keen. New Brunswick, N.J.: Rutgers University Press, 1992.

Competitive Enterprise Institute. "We Call It *Life*." Available at http://www.cei.org/pages/co2.cfm.

Coniff, Richard. "Counting Carbons." *Discover*, August 2005, 54–61.

Costa, Marcos Heil, Aurélie Botta, and Jeffrey A. Cardille. "Effects of Large-Scale Changes in Land Cover on the Discharge of the Tocantins River, Southeastern Amazonia." *Journal of Hydrology* 283 (2003): 206–217. Available at http://www.sage.wisc.edu/pubs/articles/A-E/Costa/Costa2003JHydro.pdf.

Costanza, Robert, et al. "The Value of the World's Ecosystem Services and Natural Capital." *Nature* 387 (1997): 253–260.

Cowles, Henry Chandler. "The Ecological Relations of the Vegetation on the Sand Dunes of Lake Michigan." *Botanical Gazette* 27 (1899): 95–117, 167–202, 281–308, 361–391.

Cox, T. S., et al. "Breeding Perennial Grain Crops." *Critical Reviews in Plant Sciences* 21 (2002): 59–91.

———. "Prospects for Developing Perennial Grains." *BioScience* 56 (2006): 649–659. Available at http://www.landinstitute.org/pages/Bioscience_PerennialGrains.pdf.

Crews, T. E. "Perennial Crops and Endogenous Nutrient Supplies." *Renewable Agriculture and Food Systems* 20 (2005): 25–37.

Crick, H.Q.P., et al. "UK Birds Are Laying Eggs Earlier." *Nature* 388 (1997): 526.

Crosby, Alfred W. *Ecological Imperialism: The Ecological Expansion of Europe, 900–1900*. New York: Cambridge University Press, 1986.

Dale, Virginia H., Frederick J. Swanson, and Charles M. Crisafulli, eds. *Ecological Responses to the 1980 Eruption of Mount St. Helens*. New York: Springer, 2005.

Dauvergne, Peter. *Shadows in the Forest: Japan and the Politics of Timber in Southeast Asia*. Cambridge, Mass.: MIT Press, 1997.

Davidson, Eric A. *You Can't Eat GNP: Economics as if Ecology Mattered*. Cambridge, Mass.: Perseus, 2000.

Dean, Cornelia. "The Preservation Predicament." *New York Times*, January 29, 2008.

Dean, Jeffrey S. "Anthropogenic Environmental Change in the Southwest as Viewed from the Colorado Plateau." Chapter 10 in Charles L. Redman, Steven R. James, Paul R. Fish, and J. Daniel Rogers, eds., *The Archaeology of Global Change: The Impact of Humans on Their Environment*. Washington, D.C.: Smithsonian Institution Press, 2004.

Diamond, Jared. *Collapse: How Societies Choose to Fail or Succeed*. New York: Viking, 2005.

———. "Easter Island Revisited." *Science* 317 (2007): 1692–1694.

———. *Guns, Germs, and Steel: The Fates of Human Societies*. New York: Norton, 1997.

Doebley, J., A. Stec, and L. Hubbard. "The Evolution of Apical Dominance in Maize." *Nature* 386 (1997): 485–488.

Douglas, Ellen M., et al. "Policy Implications of a Pan-Tropic Assessment of the Simultaneous Hydrological and Biodiversity Impacts of Deforestation." In Eric Craswell et al., eds., *Integrated Assessment of Water Resources and Global Change: A North-South Analysis*, 211–232. Dordrecht: SpringerNetherlands, 2007.

Dow, Kristin, and Thomas E. Downing. *The Atlas of Climate Change*. Berkeley: University of California Press, 2006.

Dregne, H. E., and N-T. Chou. "Global Desertification Dimensions and Costs." In H. E. Dregne, ed., *Degradation and Restoration of Arid Lands*. Lubbock: Texas Tech University Press, 1992.

Dunnett, Nigel, and Noel Kingsbury. *Planting Green Roofs and Living Walls*. Portland, Ore.: Timber Press, 2004.

Elias, Thomas, and Peter Dykeman. *Edible Wild Plants: A North American Field Guide*. New York: Sterling, 1990.

Energy Action Coalition. "Power Shift 2007." Available at http://www.powershift07.org.

Energy Information Administration. "Environment: Energy-Related Emissions Data and Environmental Analyses." Available at http://www.eia.doe.gov/environment.html.

Environmental Energy Technologies, Lawrence Berkeley Laboratory. "Vegetation." Available at http://eetd.lbl.gov/HeatIsland/Vegetation.

Erwin, Douglas H. *Extinction: How Life on Earth Nearly Ended 250 Million Years Ago*. Princeton, N.J.: Princeton University Press, 2006.

Fagan, Brian. *The Long Summer: How Climate Changed Civilization*. New York: Basic Books, 2004.

Fall, Patricia L., et al. "Environmental Impacts of the Rise of Civilization in the Southern Levant." Chapter 7 in Charles L. Redman, Steven R. James, Paul R. Fish, and J. Daniel Rogers, eds., *The Archaeology of Global Change: The Impact of Humans on Their Environment*. Washington, D.C.: Smithsonian Institution Press, 2004.

Fargione, Joseph, et al. "Land Clearing and the Biofuel Carbon Debt." *Science* 319 (2008): 1235–1238.

Ferguson, Rob. *The Devil and the Disappearing Sea: A True Story about the Aral Sea Catastrophe*. Vancouver: Raincoast Books, 2005.

Field, Christopher B., Michael J. Behrenfeld, James T. Randerson, and Paul Falkowski. "Primary Production of the Biosphere: Integrating Terrestrial and Oceanic Components." *Science* 281 (1998): 237–240.

Fischetti, Mark. "Living Cover." *Scientific American*, May 2008, 104–105.

Fish, Suzanne K., and Paul R. Fish. "Unsuspected Magnitudes: Expanding the Scale of Hohokam Agriculture." Chapter 11 in Charles L. Redman, Steven R. James, Paul R. Fish, and J. Daniel Rogers, eds., *The Archaeology of Global Change: The Impact of Humans on Their Environment*. Washington, D.C.: Smithsonian Institution Press, 2004.

Fitter, A. H., and R.S.R. Fitter. "Rapid Changes in Flowering Time in British Plants." *Science* 296 (2002): 1689–1691.

Flannery, Tim. *The Weather Makers: How Man Is Changing the Climate and What It Means for Life on Earth*. New York: Atlantic Monthly Press, 2005.

Foster, Nelson, and Linda S. Cordell, eds. *Chilies to Chocolate: Food the Americas Gave the World*. Tucson: University of Arizona Press, 1992.

Fox, Douglas. "Back to the No-Analog Future?" *Science* 316 (2007): 823–825.

Francaviglia, Richard V. *The Cast Iron Forest: A Natural and Cultural History of the North American Cross Timbers*. Austin: University of Texas Press, 2000.

Fu, Qiang, et al. "Enhanced Mid-Latitude Tropospheric Warming in Satellite Measurements." *Science* 312 (2006): 1179.

Gardner, Timothy. "Big Investors Urge U.S. to Slash CO_2 Emissions." Environmental News Network, March 20, 2007. Available at http://www.enn.com.

Gaulin, Mark. "The Green Building Advantage—Green Roofs Are the Way to Grow." Available at http://www.greenroof.com/image_body/pdf/17-18_Reprint_02.pdf.

Bibliography

Geisler, Norman. *Knowing the Truth about Creation: How It Happened and What It Means for Us*. Ann Arbor, Mich.: Servant Books, 1989.

Gibson, J. Phil, Stanley A. Rice, and Clare M. Stucke. "Comparison of population genetic diversity between a rare, narrowly distributed species and a common, widespread species of *Alnus* (Betulaceae)." *American Journal of Botany* 95 (2008): 588–596.

Gill, David S., and Peter L. Marks. "Tree and Shrub Seedling Colonization of Old Fields in Central New York." *Ecological Monographs* 61 (1991): 183–205.

Glantz, Michael H., ed. *Drought Follows the Plow*. New York: Cambridge University Press, 1994.

Global Footprint Network. "National Footprints." Available at http://www.footprintnetwork.org/gfn_sub.php?content=national_footprints.

Glover, Jerry D., Cindy M. Cox, and John P. Reganold. "Future Farming: A Return to Roots?" *Scientific American*, August 2007, 82–89.

Goddard Space Flight Center, Scientific Visualization Studio. "NASA Satellite Confirms Urban Heat Islands Increase Rainfall around Cities." Available at http://www.gsfc.nasa.gov/goddardnews/20020621/urbanrain.html.

———. "Urban Growth Seen from Space." Available at http://svs.gsfc.nasa.gov/stories/AAAS.

Gordon, Doria R., and Kevin J. Rice. "Competitive suppression of *Quercus douglasii* (Fagaceae) seedling emergence and growth." *American Journal of Botany* 87 (2000): 986–994.

Gordon, Line J. "Land cover change and water vapour flows: Learning from Australia." *Philosophical Transactions of the Royal Society of London B (Biology)* 358 (2003): 1973–1984.

Gordon, Line J., et al. "Human modification of global water vapor flows from the land surface." *Proceedings of the National Academy of Sciences* 102 (2005): 7612–7617.

Gore, Al. *Earth in the Balance: Ecology and the Human Spirit*. Boston: Houghton Mifflin, 1992.

———. *An Inconvenient Truth: The Planetary Emergency of Global Warming and What We Can Do about It*. New York: Rodale, 2006.

Gorsevski, Virginia, Haider Taha, Dale Quattrochi, and Jeff Luval. "Air Pollution Prevention through Urban Heat Island Mitigation: An Update on the Urban Heat Island Pilot Project." Available at http://www.ghcc.msfc.nasa.gov/uhipp/epa_doc.pdf.

Gorski, Eric. "Southern Baptists Question Human Role in Global Warming; Say Regulations Hurt Poor." Environmental News Network, June 14, 2007. Available at http://www.enn.com.

Goswami, B. N., et al. "Increasing Trend of Extreme Rain Events over India in a Warming Environment." *Science* 314 (2006): 1442–1445.

Grabherr, G., M. Gottfried, and H. Pauli. "Climate Effects on Mountain Plants." Nature 369 (1994): 448.

Gram, Dave. "Judge Rejects Carmakers' Emission Suit." Environmental News Network, September 12, 2007. Available at http://www.enn.com.

Green Belt Movement. "The Green Belt Movement." Available at http://www.greenbeltmovement.org.

Greenroofs.com. "Greenroof Projects Database." Available at http://www.greenroofs.com/projects/plist.php.

Grogan, P., et al. "Below-Ground Ectomycorrhizal Community Structure in a Recently Burned Bishop Pine Forest." *Journal of Ecology* 88 (2000): 1051–1062.

Guevara, Sergio, et al. "The Role of Remnant Forest Trees in Tropical Secondary Succession." *Plant Ecology* 66 (1986): 77–84.

Gullison, Raymond E., et al. "Tropical Forests and Climate Policy." *Science* 316 (2007): 985–986.

Gutro, Rob. "There's a Change in Rain around Desert Cities." Earth Observatory, National Aeronautics and Space Administration. Available at http://earthobservatory.nasa.gov//Newsroom/NasaNews/2006/2006072522726.html.

Haines, David B. "The Value of Trees in City of New Berlin." Available at http://gis.esri.com/library/userconf/proc01/professional/papers/pap1021/p1021.htm.

Halweil, Brian, and Danielle Nierenberg. "Meat and Seafood: The Global Diet's Most Costly Ingredients." Chapter 5 in Worldwatch Institute, *State of the World: Innovations for a Sustainable Economy.* New York: Norton, 2008.

"Hansen Files Landmark Bill to Restore Original Intent of Endangered Species Act." *Sierra Times,* November 13, 2002. Available at http://www.sierratimes.com/02/11/14/arst111402.htm.

Hargreaves, Steve. "Exxon Linked to Climate Change Pay Out: Think Tank Offers Scientists $10,000 to Criticize UN Study Confirming Global Warming and Placing Blame on Humans." *CNN Money,* February 5, 2007. Available at http://money.cnn.com/2007/02/02/news/companies/exxon_science/index.htm.

Harlan, Jack R. *The Living Fields: Our Agricultural Heritage.* New York: Cambridge University Press, 1994.

Heartland Institute. "Heartland Institute." Available at http://www.heartland.org.

Heath, James, et al. "Rising Atmospheric CO_2 Reduces Sequestration of Root-Derived Soil Carbon." *Science* 309 (2005): 1711–1713.

Hebert, H. Josef. "EPA Scientists Complain about Political Pressure." Environmental News Network, April 23, 2008. Available at http://www.enn.com.

———. "House Approves Fees, Taxes on Oil Companies; Plans to Use Money for Renewable Fuels." Environmental News Network, January 19, 2007. Available at http://www.enn.com.

———. "White House Official Singled Out for Editing Climate Reports to Work for ExxonMobil." Environmental News Network, June 15, 2005. Available at http://www.enn.com.

Hecht, Susanna, and Alexander Cockburn. *The Fate of the Forest: Developers, Destroyers, and Defenders of the Amazon.* New York: HarperCollins, 1989.

Heiser, Charles B. *Seed to Civilization: The Story of Food.* Cambridge, Mass.: Harvard University Press, 1973.

Hill, Michael. "Do Trees Make It O.K. to Drive an SUV?" Environmental News Network, May 28, 2007. Available at http://www.enn.com.

Hillel, Daniel. *Out of the Earth: Civilization and the Life of the Soil.* Berkeley: University of California Press, 1992.

Hinman, C. Wiley, and Jack W. Hinman. *The Plight and Promise of Arid Land Agriculture.* New York: Columbia University Press, 1992.

Hobhouse, Henry. *Seeds of Change: Five Plants That Transformed Mankind.* New York: Harper and Row, 1987.

Holl, Karen D. "Effect of Shrubs on Tree Seedling Establishment in an Abandoned Tropical Pasture." *Journal of Ecology* 90 (2002): 179–187.

Holl, Karen D., et al. "Tropical Montane Forest Restoration in Costa Rica: Overcoming Barriers to Dispersal and Establishment." *Restoration Ecology* 8 (2000): 339–349.

Hopfensperger, Kristine N. "A Review of Similarity between Seed Bank and Standing Vegetation across Ecosystems." *Oikos* 116 (2007): 1438–1448.

Hopkinson, Francis. *The Miscellaneous Essays and Occasional Writings of Francis* Hopkinson. Philadelphia, n.d.

Howat, Ian M., Ian Joughin, and Ted A. Scambos. "Rapid Changes in Ice Discharge from Greenland Outlet Glaciers." *Science* 315 (2007): 1559–1561.

Hoyos, C. D., et al. "Deconvolution of the Factors Contributing to the Increase in Global Hurricane Intensity." *Science* 312 (2006): 94–97.

Huey, Raymond B., and Peter D. Ward. "Hypoxia, Global Warming, and Terrestrial Late Permian Extinctions." *Science* 308 (2005): 398–401.

Hunt, Terry L. "Rethinking the Fall of Easter Island." *American Scientist* 94 (2006).

Huxley, Anthony. *Green Inheritance: Saving the Plants of the World*. Berkeley: University of California Press, 2006.

Intergovernmental Panel on Climate Change. *Climate Change 2007: Climate Change Impacts, Adaptation, and Vulnerability. Summary for Policymakers*. Available at http://www.ipcc.ch/SPM6avr07.pdf.

———. *Climate Change 2007: Mitigation. Summary for Policymakers*. Available at http://www.ipcc.ch/SPM040507.pdf.

———. *Climate Change 2007: The Physical Science Basis. Summary for Policymakers*. Available at http://www.ipcc.ch/SPM2feb07.pdf.

———. *Climate Change 2007: Synthesis Report. Summary for Policymakers*. Available at http://www.ipcc.ch/pdf/assessment-report/ar4/syr/ar4_syr_spm.pdf.

Iverson, Louis R., and Anantha M. Prasad. "Potential Redistribution of Tree Species Habitats under Five Climate Change Scenarios in the Eastern U.S." *Forest Ecology and Management* 155 (2002): 205–222.

———. "Predicting Abundance of Eighty Tree Species Following Climate Change in the Eastern United States." *Ecological Monographs* 68 (1998): 465–485.

Jackson, Wes. "Natural Systems Agriculture: A Radical Alternative." *Agriculture, Ecosystems and Environment* 88 (2002): 111–117. Available at http://www.landinstitute.org/vnews/display.v/ART/2001/04/17/3aa80bec9.

———. *New Roots for Agriculture*. Lincoln: University of Nebraska Press, 1980.

Jacobson, Mark, and Gilbert M. Masters. "Exploiting Wind versus Coal." Science 293 (2001): 1438.

Jacobson, Thorkild, and Robert M. Adams. "Salt and Silt in Ancient Mesopotamian Agriculture: Progressive Changes in Soil Salinity and Sedimentation Contributed to the Breakup of Past Civilizations." *Science* 128 (1958): 1251–1258.

Jones, P. D., and Michael E. Mann. "Climate over Past Millennia." *Reviews of Geophysics* 42 (2004): RG2002, doi:10.1029/2003RG000143.

Keeley, Jon E., and C. D. Fotheringam. "Smoke-Induced Seed Germination in Californian Chaparral." *Science* 276 (1997): 1248–1250.

Keeley, Jon E., and Sterling C. Keeley. "Chaparral." Chapter 6 in Michael G. Barbour and William Dwight Billings, eds., *North American Terrestrial Vegetation* (New York: Cambridge University Press, 1988).

———. "Demographic Structure of California Chaparral in the Long-Term Absence of Fire." *Journal of Vegetation Science* 3 (1992): 79–90.

Kerr, Richard A. "Global Warming: How Urgent Is Climate Change?" Science 318 (2007): 1230–1231.

———. "Global Warming Is Changing the World." *Science* 316 (2007): 188–190.

———. "Is Battered Arctic Sea Ice down for the Count?" *Science* 318 (2007): 33–34.

———. "Mother Nature Cools the Greenhouse, but Hotter Times Still Lie Ahead." *Science* 320 (2008): 595.

———. "Record U.S. Warmth of 2006 Was Part Natural, Part Greenhouse." *Science* 317 (2007): 182–183.

———. "A Worrying Trend of Less Ice, Higher Seas." *Science* 311 (2006): 1698–1701.

King, Anthony W. "Plant Respiration in a Warmer World." *Science* 312 (2006): 536–537.

Kingsolver, Barbara. *Small Wonder*. New York: HarperCollins, 2002.

Kintisch, Eli. "Industry Conservation Programs Face White House Cuts." *Science* 312 (2006): 675.

———. "U.S. Supreme Court Gets Arguments for EPA to Regulate CO_2." *Science* 313 (2006): 1375.

Kirch, Patrick V. "Oceanic Islands: Microcosms of Global Change." Chapter 1 in Charles L. Redman, Steven R. James, Paul R. Fish, and J. Daniel Rogers, eds., *The Archaeology of Global Change: The Impact of Humans on Their Environment*. Washington, D.C.: Smithsonian Institution Press, 2004.

Kleidon, A., K. Fraedrich, and M. Heimann. "A Green Planet versus a Desert World: Estimating the Maximum Effect of Vegetation on the Land Surface Climate." *Climate Change* 44 (2000): 417–493.

Knepp, R. E., et al. "Elevated CO_2 Reduces Leaf Damage by Insect Herbivores in a Forest Community." *New Phytologist* 167 (2005): 207–218.

Koeppel, Dan. *Banana: The Fate of the Fruit That Changed the World*. New York: Hudson Street Press, 2007.

Kohler, Timothy A. "Pre-Hispanic Human Impact on Upland North American Southwestern Environments: Evolutionary Ecological Perspectives." Chapter 12 in Charles L. Redman, Steven R. James, Paul R. Fish, and J. Daniel Rogers, eds., *The Archaeology of Global Change: The Impact of Humans on Their Environment*. Washington, D.C.: Smithsonian Institution Press, 2004.

Kolbert, Elizabeth. "The Curse of Akkad." Chapter 5 in *Field Notes from a Catastrophe: Man, Nature, and Climate Change*. New York: Bloomsbury, 2006.

Körner, Christian, et al. "Carbon Flux and Growth in Mature Deciduous Forest Trees Exposed to Elevated CO_2." *Science* 309 (2005): 1360–1362.

———. "Small Differences in Arrival Time Influence Composition and Productivity of Plant Communities." *New Phytologist* 177 (2008): 698–705.

Krupp, Fred, and Miriam Horn. *Earth, the Sequel: The Race to Reinvent Energy and Stop Global Warming*. New York: Norton, 2008.

Kugler, Sara. "Sixteen Cities to Go Green under Clinton Plan." Environmental News Network, May 16, 2007. Available at http://www.enn.com.

Kunzig, Robert. "Drying of the West." *National Geographic*, February 2008, 90–113.

L'Abbé, Sonnet. "Green Roofs in Winter: Hot Design for a Cold Climate." Available at http://www.news.utoronto.ca/bin6/051117-1822.asp.

Lewington, Anna. *Plants for People*. New York: Oxford University Press, 1990.

Lewington, Anna, and Edward Parker. *Ancient Trees: Trees That Live for a Thousand Years*. London: Collins and Brown, 1999.

Lewis, Charles A. *Green Nature, Human Nature: The Meaning of Plants in Our Lives*. Urbana: University of Illinois Press, 1996.

Lichter, John. "Colonization Constraints during Primary Succession on Coastal Lake Michigan Sand Dunes." *Journal of Ecology* 88 (2000): 825–839.

Likens, Gene E., et al. "Effects of Forest Cutting and Herbicide Treatment on Nutrient Budgets in the Hubbard Brook Watershed-Ecosystem." *Ecological Monographs* 40 (1970): 23–47. Data available at http://tiee.ecoed.net/vol/v1/data_sets/hubbard/hubbard_faculty.xls.

Lindeman, Ray L. "The Trophic-Dynamic Aspect of Ecology." *Ecology* 23 (1942): 399–418.

Bibliography

Linden, Eugene. *The Winds of Change: Climate, Weather, and the Destruction of Civilizations*. New York: Simon and Schuster, 2007.

Little, Charles E. *The Dying of the Trees: The Pandemic in America's Forests*. New York: Penguin, 1995.

Lobell, David B., et al. "Prioritizing Climate Change Adaptation Needs for Food Security in 2030." *Science* 319 (2008): 607–610. Reviewed by Brown, Molly E., and Christopher C. Funk. "Food Security under Climate Change." *Science* 319 (2008): 580–581.

LocalHarvest, Inc. "LocalHarvest: Real Food, Real Farmers, Real Community." Available at http://www.localharvest.org.

Long, Colleen. "System Relies on Ice to Chill Buildings." Environmental News Network, July 16, 2007. Available at http://www.enn.com.

Long, Stephen P., et al. "Food for Thought: Lower-Than-Expected Crop Yield Stimulation with Rising CO_2 Concentrations." *Science* 312 (2006): 1918–1921.

Looney, J. Jefferson, ed. *Papers of Thomas Jefferson, Retirement Series*, vol. 4. Princeton, N.J.: Princeton University Press, 2007.

Lovejoy, Thomas E., and Lee Hannah, eds. *Climate Change and Biodiversity*. New Haven, Conn.: Yale University Press, 2005.

Lovell, Jeremy. "Britain Proposes Legal Limits on Carbon Emissions." Environmental News Network, March 14, 2007. Available at http://www.enn.com.

Lowman, Margaret. *Life in the Treetops: Adventures of a Woman in Field Biology*. New Haven, Conn.: Yale University Press, 2000.

Luthcke, S. B., et al. "Recent Greenland Ice Mass Loss by Drainage System from Satellite Gravity Observations." *Science* 314 (2006): 1286–1289.

Lynas, Mark. *Six Degrees: Our Future on a Hotter Planet*. Washington, D.C.: National Geographic, 2008.

Maathai, Wangari. *The Green Belt Movement: Sharing the Approach and the Experience*. New York: Lantern Books, 2003.

———. *Unbowed: A Memoir*. New York: Random House, 2006.

Malhi, Yavinder, et al. "Climate Change, Deforestation, and the Fate of the Amazon." *Science* 319 (2008): 169–172.

Mann, Charles C. *1491: New Revelations of the Americas before Columbus*. New York: Knopf, 2005.

Mann, Michael E., R. S. Bradley, and M. K. Hughes. "Global-Scale Temperature Patterns and Climate Forcing over the Past Six Centuries." *Nature* 392 (1998): 779–787.

Matossian, Mary Kilbourne. *Poisons of the Past: Molds, Epidemics, and History*. New Haven, Conn.: Yale University Press, 1991.

Maxwell, Nicole. *Witch-Doctor's Apprentice: Hunting for Medicinal Plants in the Amazon* (New York: Citadel, 1961).

McEachern, George Ray. "A Texas Grape and Wine History." Available at http://aggie-horticulture.tamu.edu/southerngarden/Texaswine.html.

McKenney, Daniel W., et al. "Potential Impacts of Climate Change on the Distribution of North American Trees." *BioScience* 57 (2007): 939–948.

McKibben, Bill. *Deep Economy: The Wealth of Communities and the Durable Future*. New York: Henry Holt, 2007.

McKibben, Bill. *The End of Nature*. New York: Random House, 2006.

———. "It's Easy Being Green." *Mother Jones*, July/August 2002.

Meier, Mark F., et al. "Glaciers Dominate Eustatic Sea-Level Rise in the 21st Century." *Science* 317 (2007): 1064–1067.

Merwin, W. S. *Migration: New and Selected Poems*. Port Townsend, Wash.: Copper Canyon Press 2007.

Micklin, Philip, and Nikolay V. Aladin. "Reclaiming the Aral Sea." *Scientific American*, April 2008, 64–71.

Miller, Naomi F. "Long-Term Vegetation Changes in the Near East." Chapter 6 in Charles L. Redman, Steven R. James, Paul R. Fish, and J. Daniel Rogers, eds., *The Archaeology of Global Change: The Impact of Humans on Their Environment*. Washington, D.C.: Smithsonian Institution Press, 2004.

Miller-Rushing, Abraham J., et al. "Photographs and Herbarium Specimens as Tools to Document Phenological Changes in Response to Global Warming." *American Journal of Botany* 93 (2006): 1667–1674.

Mills, Evan. "Insurance in a Climate of Change." Science 309 (2005): 1040–1044.

Mintz, Sidney W. *Sweetness and Power: The Place of Sugar in Modern History*. New York: Penguin, 1986.

Mishchenko, Michael I., et al. "Long-Term Satellite Record Reveals Likely Recent Aerosol Trend." *Science* 315 (2007): 1543.

Miura, S., et al. "Protective Effect of Floor Cover against Soil Erosion on Steep Slopes Forested with *Chamaecyparis obtusa* (Hinoki) and Other Species." *Journal of Forest Research* 8 (2003): 23–75.

Montañez, Isabel P., et al. "CO_2-Forced Climate and Vegetation Instability during Late Paleozoic Deglaciation." *Science* 315 (2007): 87–91.

Montgomery, David R. *Dirt: The Erosion of Civilizations*. Berkeley: University of California Press, 2007.

Mooney, Chris. *The Republican War on Science*. New York: Basic Books, 2005.

Musil, Robert K. *Hope for a Heated Planet: How Americans Are Fighting Global Warming and Building a Better Future*. New Brunswick, N.J.: Rutgers University Press, forthcoming.

Myneni, R. B., et al. "Increased Plant Growth in the Northern High Latitudes from 1981 to 1991." *Nature* 386 (1997): 698–702. See also commentary by I. Fung, "A Greener North," *Nature* 386 (1997): 659–660.

Nabhan, Gary Paul. *Why Some Like It Hot: Food, Genes, and Cultural Diversity*. Washington, D.C.: Island Press, 2004.

Nadkarni, Nalini M. *Between Earth and Sky: Our Intimate Connections to Trees*. Berkeley: University of California Press, 2008.

Nair, U. S., et al. "Impact of Land Use on Costa Rican Tropical Montane Cloud Forests: Sensitivity of Cumulus Cloud Field Characteristics to Lowland Deforestation." *Journal of Geophysical Research* 108 (2003): 4206.

Naranjo, Laura. "Seeing the City for the Trees." Earth Science Data and Services, National Aeronautics and Space Administration. Available at http://nasadaacs.eos.nasa.gov/articles/2004/2004_urban.html.

National Oceanic and Atmospheric Administration. "A Paleo Perspective on Global Warming: The Instrumental Data of Past Global Temperatures." Available at http://www.ncdc.noaa.gov/paleo/globalwarming/instrumental.html.

———. "A Paleo Perspective on Global Warming: Paleoclimatic Data for the Last 2000 Years." Available at http://www.ncdc.noaa.gov/paleo/globalwarming/paleolast.html.

———. "Trends in Carbon Dioxide." Available at http://www.esrl.noaa.gov/gmd/ccgg/trends/co2_data_mlo.html.

National Research Council, Board on Science and Technology for International Development. *Lost Crops of Africa. Volume 1: Grains*. Washington, D.C.: National Academy Press, 1996.

Bibliography

National Sustainable Agriculture Information Service. "Agroforestry Overview." Available at http://www.attra.org/attra-pub/agroforestry.html.

Normile, Dennis. "Getting at the Roots of Killer Dust Storms." *Science* 317 (2007): 314–316.

Noticewala, Sonal. "At Australia's Bunny Fence, Variable Cloudiness Prompts Climate Study." *New York Times*, August 14, 2007.

Nussey, Daniel H., et al. "Selection on Heritable Phenotypic Plasticity in a Wild Bird Population." *Science* 310 (2005): 304–306.

O'Hara, Sarah L., and Sarah E. Metcalfe. "Late Holocene Environmental Change in West-Central Mexico." Chapter 8 in Charles L. Redman, Steven R. James, Paul R. Fish, and J. Daniel Rogers, eds., *The Archaeology of Global Change: The Impact of Humans on Their Environment*, Washington, D.C.: Smithsonian Institution Press, 2004.

Oki, Taikan, and Shinjiro Kanae. "Global Hydrological Cycles and World Water Resources." *Science* 313 (2006): 1068–1072.

Oreskes, Naomi. "The Scientific Consensus on Climate Change." *Science* 306 (2004): 1686.

Ortuäno, M., J. Alarcâon, E. Nicolâas, and A. Torrecillas. "Water Status Indicators of Lemon Trees in Response to Flooding and Recovery." *Biologia Plantarum* 51 (2007): 292–296.

Osborn, Timothy J., and Keith R. Briffa. "The Spatial Extent of 20th-Century Warmth in the Context of the Past 1200 Years." *Science* 311 (2006): 841–844.

Otto-Bliesner, Bette L., et al. "Simulating Arctic Climate Warmth and Icefield Retreat in the Last Interglaciation." *Science* 311 (2006): 1751–1753.

Overpeck, Jonathan T., et al. "Paleoclimatic Evidence for Future Ice-Sheet Instability and Rapid Sea-Level Rise." *Science* 311 (2006): 1747–1750.

Pagani, Mark, et al. "An Ancient Carbon Mystery." *Science* 314 (2006): 1556–1557.

———. "Marked Decline in Atmospheric Carbon Dioxide Concentrations during the Paleogene." *Science* 309 (2005): 600–603.

Pakenham, Thomas. *Remarkable Trees of the World*. New York: Norton, 2002.

Parmesan, Camille, and Gary Yohe. "A Globally Coherent Fingerprint of Climate Change Impacts across Natural Systems." *Nature* 421 (2003): 37–42.

———. "Poleward Shifts in Geographical Ranges of Butterfly Species Associated with Regional Warming." *Nature* 399 (1999): 579–583.

Pearce, Fred. *With Speed and Violence: Why Scientists Fear Tipping Points in Climate Change.* Boston: Beacon, 2007.

Perlin, John. *A Forest Journey: The Role of Wood in the Development of Civilization*. Cambridge, Mass.: Harvard University Press, 1989.

Peterson, David L., Edward G. Schreiner, and Nelsa M. Buckingham. "Gradients, Vegetation and Climate: Spatial and Temporal Dynamics in the Olympic Mountains, U.S.A." *Global Ecology and Biogeography Letters* 6 (1997): 7–17.

Petit, J. R., et al. "Historical Isotopic Temperature Record from the Vostok Ice Core." In *Trends: A Compendium of Data on Global Change*, Carbon Dioxide Information Analysis Center, Oak Ridge National Laboratory, 2000. Available at http://cdiac.ornl.gov/trends/temp/vostok/jouz_tem.htm.

Pew Center on Global Climate Change. "Pew Center on Global Climate Change." Available at http://www.pewclimate.org.

Phillips, Kevin. *American Theocracy: The Peril and Politics of Radical Religion, Oil, and Borrowed Money in the 21st Century*. New York: Penguin, 2007.

Pickett, Steward T. A. "Population Patterns through Twenty Years of Oldfield Succession." *Plant Ecology* 49 (1982): 45–59.

———. "Succession: An Evolutionary Interpretation." *American Naturalist* 110 (1976): 107–119.

Pielou, E. C. *After the Ice Age: Return of Life to Glaciated North America.* Chicago: University of Chicago Press, 1992.

Pimentel, David. "Livestock Production: Energy Inputs and the Environment." Canadian *Society of Animal Science Proceedings* 47 (1997): 17–26.

Pimentel, David, and Marcia Pimentel. "Sustainability of Meat-Based and Plant-Based Diets and the Environment." *American Journal of Clinical Nutrition* 78 (2003): 660–663 (supplement). Available at http://www.ajcn.org/cgi/content/full/78/3/660S?ck=nck.

Plato. *Timaeus and Critias.* New York: Penguin, 1972.

Plotkin, Mark. *Tales of a Shaman's Apprentice: An Ethnobotanist Searches for New Medicines in the Amazon Rain Forest.* New York: Viking, 1993.

Plug In America. "Plug In America." Available at http://www.pluginamerica.com.

Pollan, Michael. *The Botany of Desire: A Plant's-Eye View of the World.* New York: Random House, 2001.

———. *The Omnivore's Dilemma: A Natural History of Four Meals.* New York: Penguin, 2006.

Ponting, Clive. *A Green History of the World: The Environment and the Collapse of Great Civilizations.* New York: Penguin, 1991.

Pope, Carl, and Paul Rauber. *Strategic Ignorance: Why the Bush Administration Is Recklessly Destroying a Century of Environmental Progress.* Berkeley: University of California Press, 2004.

Postel, Sandra. *Pillar of Sand: Can the Irrigation Miracle Last?* New York: Norton, 1999.

Rahmsdorf, Stefan, et al. "Recent Climate Observations Compared to Projections." *Science* 316 (2007): 709.

Randall, David K. "Perhaps Only God Can Make a Tree, but Only People Could Put a Dollar Value on It." *New York Times,* April 18, 2007.

RealClimate. "RealClimate: Climate Science from Climate Scientists." Available at http://www.realclimate.org.

Redman, Charles L. "Environmental Degradation and Early Mesopotamian Civilization." Chapter 8 in Charles L. Redman, Steven R. James, Paul R. Fish, and J. Daniel Rogers, eds., *The Archaeology of Global Change: The Impact of Humans on Their Environment.* Washington, D.C.: Smithsonian Institution Press, 2004.

———. *Human Impact on Ancient Environments.* Tucson: University of Arizona Press, 1999.

Redman, Charles L., Steven R. James, Paul R. Fish, and J. Daniel Rogers, eds., The *Archaeology of Global Change: The Impact of Humans on Their Environment.* Washington, D.C.: Smithsonian Institution Press, 2004.

Retallack, Gregory J. "A 300-Million-Year Record of Atmospheric Carbon Dioxide from Fossil Cuticles." *Nature* 411 (2001): 287–290.

Revkin, Andrew C. "Scientists Report Severe Retreat of Arctic Ice." *New York Times,* September 21, 2007.

———. "A Young Bush Appointee Resigns His Post at NASA." *New York Times,* February 8, 2006.

Rice, Stanley A. "Agriculture, Origin of." In *Encyclopedia of Evolution.* New York: Facts on File, 2006.

———. "Speciation." In *Encyclopedia of Evolution.* New York: Facts on File, 2006.

Rice, Stanley A., and John McArthur. "Water Flow through Xylem: An Investigation of a Fluid Dynamics Principle Applied to Plants." *American Biology Teacher* 66 (2004): 203–210.

Bibliography

Righelato, Renton, and Dominick V. Spracklen. "Carbon Mitigation by Biofuels or by Saving and Restoring Forests?" *Science* 317 (2007): 902.

Rindos, David. *The Origins of Agriculture: An Evolutionary Perspective.* San Diego: Academic Press, 1984.

Rocky Mountain Institute. "Profitable Innovations for Energy and Resource Efficiency." Available at http://www.rmi.org.

Rogers, Christine A., et al., "Interaction of the Onset of Spring and Elevated Atmospheric CO_2 on Ragweed (*Ambrosia artemesiifolia* L.) Pollen Production." *Environmental Health Perspectives* 114 (2006): 865–869.

Romm, Joseph. *Hell and High Water: Global Warming—The Solution and the Politics—And What We Should Do.* New York: William Morrow, 2006.

Rosenthal, Elisabeth. "Near Arctic, Seed Vault Is a Fort Knox of Food." *New York Times,* February 29, 2008.

———. "Vatican Tree Penance: Forgive Us Our CO_2." *New York Times,* September 17, 2007.

Royte, Elizabeth. "The Gospel According to John." *Discover,* February 2001. Available at http://discovermagazine.com/2001/feb/featgospel.

Ruddiman, William F. *Plows, Plagues, and Petroleum: How Humans Took Control of Climate.* Princeton, N.J.: Princeton University Press, 2005.

Salaman, Redcliffe N. *The History and Social Influence of the Potato.* Cambridge: Cambridge University Press, 1949.

Sampson, D. A., J. M. Vose, and H. L. Allen. "A Conceptual Approach to Stand Management Using Leaf Area Index as the Integral of Site Structure, Physiological Function, and Resource Supply." In *Ninth Biennial Southern Silvicultural Research Conference,* USDA Forest Service, Southern Research Station General Technical Report SRS-20, 447–451. Available at http://cwt33.ecology.uga.edu/publications/1233.pdf.

Sanz-Elorza, Mario, et al. "Changes in the High-Mountain Vegetation of the Central Iberian Peninsula as a Probable Sign of Global Warming." *Annals of Botany* 92 (2003): 273–280.

Schlesinger, William H. "Global Change Ecology." *Trends in Ecology and Evolution* 21 (2006): 348–351.

Schlosser, Eric. *Fast Food Nation.* New York: Harper Perennial, 2005.

Schoups, Gerrit, et al. "Sustainability of Irrigated Agriculture in the San Joaquin Valley, California." *Proceedings of the National Academy of Sciences USA* 102 (2005): 15352–15356.

Schrader, James A., William R. Graves, Stanley A. Rice, and J. Phil Gibson. "Differences in Shade Tolerance Help Explain Varying Success in Two Sympatric *Alnus* Species." *International Journal of Plant Sciences* 167 (2006): 979–989.

Schröter, Dagmar, et al. "Ecosystem Service Supply and Vulnerability to Global Change in Europe." *Science* 310 (2005): 1333–1337.

Schweitzer, Peter. "Offset Away Our Guilt." *USA Today,* September 25, 2007.

Scurlock, J.M.O., G. P. Asner, and S. T. Gower. *Worldwide Historical Estimates of Leaf Area Index, 1932–2000.* December 2001 technical report to the Environmental Services Division, Oak Ridge National Laboratory. Available at http://www.ornl.gov/~webworks/cppr/y2002/rpt/112600.pdf.

Seager, Richard, et al. "Model Projections of an Imminent Transition to a More Arid Climate in Southwestern North America." *Science* 316 (2007): 1181–1184.

Searchinger, Timothy, et al. "Use of U.S. Croplands for Biofuels Increases Greenhouse Gases through Emissions from Land Use Change." *Science* 319 (2008): 1238–1240.

Seed Savers Exchange. "Seed Savers Exchange." Available at http://www.seedsavers.org.

Şekercioğlu, Çağan H., et al. "Climate Change, Elevational Range Shifts, and Bird Extinctions." *Conservation Biology* 22 (2008): 140–150.

Service, Robert F. "Biofuel Researchers Prepare to Reap a New Harvest." *Science* 315 (2007): 1488–1491.

Shepherd, Andrew, and Duncan Wingham. "Recent Sea-Level Contributions of the Antarctic and Greenland Ice Sheets." *Science* 315 (2007): 1529–1532.

Shine, Keith P., and William T. Sturges. "CO$_2$ Is Not the Only Gas." *Science* 315 (2007): 1804–1805.

Shukla, J., C. Nobre, and P. Sellers. "Amazon Deforestation and Climate Change." *Science* 247 (1990): 1322–1325.

Sidle, Roy C., et al. "Erosion Processes in Steep Terrain—Truths, Myths, and Uncertainties Related to Forest Management in Southeast Asia." *Forest Ecology and Management* 224 (2006): 199–225.

Siegenthaler, Urs, et al. "Stable Carbon Cycle–Climate Relationship during the Late Pleistocene." *Science* 310 (2005): 1313–1317.

Sims, Phillip L. "Grasslands." Chapter 9 in Michael G. Barbour and William Dwight Billings, eds., *North American Terrestrial Vegetation*. New York: Cambridge University Press, 1988.

Smith, Kevin. "The Carbon Neutral Myth: Offset Indulgences for Your Climate Sins." Transnational Institute, Carbon Trade Watch. Available at http://www.carbontradewatch.org/pubs/carbon_neutral_myth.pdf.

Smith, Nigel J. H., et al. *Tropical Forests and Their Crops*. Ithaca, N.Y.: Cornell University Press, 1992.

Solbrig, Otto T., and Dorothy J. Solbrig. *So Shall You Reap: Farming and Crops in Human Affairs*. Covelo, Calif.: Island Press, 1994.

Sorkin, Andrew Ross. "A Buyout Deal That Has Many Shades of Green." *New York Times*, February 26, 2007.

Southern Exposure Seed Exchange. "Southern Exposure Seed Exchange." Available at http://www.southernexposure.com.

Spahni, Renato, et al. "Atmospheric Methane and Nitrous Oxide of the Late Pleistocene from Antarctic Ice Cores." *Science* 310 (2005): 1317–1321.

Specter, Michael. "Big Foot." *New Yorker*, February 25, 2008. Available at http://www.newyorker.com/reporting/2008/02/25/080225fa_fact_specter.

Speth, James Gustave. *Red Sky at Morning*. New Haven, Conn.: Yale University Press, 2004.

Spurlock, Morgan. *Don't Eat This Book: Fast Food and the Supersizing of America*. New York: Penguin, 2006.

Stevenson Center for Community and Economic Development. "Adlai E. Stevenson II." Available at http://www.stevensoncenter.org/stevenson.htm.

Striffler, Steve, and Mark Moberg, eds. *Banana Wars: Power, Production, and History in the Americas*. Durham, N.C.: Duke University Press, 2003.

Suttle, K. B., Meredith A. Thomsen, and Mary E. Power. "Species Interactions Reverse Grassland Responses to Changing Climate." *Science* 315 (2007): 640–642.

Sutton, Rowan T., and Daniel L. R. Hodson. "Atlantic Ocean Forcing of North American and European Summer Climate." *Science* 309 (2005): 115–118.

Tanno, Ken-Ichi, and George Willcox. "How Fast Was Wild Wheat Domesticated?" *Science* 311 (2006): 1886.

Taylor, A. Faber, Frances E. Kuo, and W. C. Sullivan. "Coping with ADD: The Surprising Connection to Green Play Settings." *Environment and Behavior* 33 (2001): 54–77.

Tennesen, Michael. "Black Gold of the Amazon." *Discover*, April 2007, 46–52.

Bibliography

Terborgh, John. *Requiem for Nature*. Covelo, Calif.: Island Press, 1999.

Thomas, Chris D., and Jack J. Lennon. "Birds Extend Their Ranges Northward." *Nature* 399 (1999): 213.

Thomas, Peter. *Trees: Their Natural History*. Cambridge: Cambridge University Press, 2001.

Tilman, David, Jason Hill, and Clarence Lehman. "Carbon-Negative Biofuels from Low-Input High-Diversity Grassland Biomass." *Science* 314 (2006): 1598–1600.

Totman, Conrad. *The Green Archipelago: Forestry in Pre-industrial Japan*. Athens: University of Ohio Press, 1998.

Trenberth, Kevin E. "Warmer Oceans, Stronger Storms." *Scientific American*, July 2007, 44–47.

Truffer, Martin, and Mark Fahnestock. "Rethinking Ice Sheet Time Scales." *Science* 315 (2007): 1508–1510.

Umina, P. A., et al. "A Rapid Shift in a Classic Clinal Pattern in *Drosophila* Reflecting Climate Change." *Science* 308 (2005): 691–693.

United Nations Environmental Programme. "The Billion Tree Campaign." Available at http://www.unep.org/billiontreecampaign.

United Nations Food and Agriculture Organization. "FAOSTAT." Available at http://faostat.fao.org/site/340/default.aspx.

———. "Terrastat." Available at http://www.fao.org/ag/agl/agll/terrastat.

United Nations Food and Agriculture Organization, and Center for International Forestry Research. "Forests and Floods: Drowning in Fiction or Thriving on Facts?" Available at ftp://ftp.fao.org/docrep/fao/008/ae929e/ae929e00.pdf.

United Nations Foundation and Sigma Xi. *Confronting Climate Change: Avoiding the Unmanageable and Managing the Unavoidable*. Scientific Expert Group Report on Climate Change and Sustainable Development. *American Scientist*, April 2007.

United States Climate Action Partnership. "USCAP: United States Climate Action Partnership." Available at http://www.us-cap.org.

U.S. Department of Agriculture. "National Agroforestry Center." Available at http://www.unl.edu/nac.

U.S. Department of Agriculture, National Agricultural Library. "Hyperaccumulators." A bibliography available at http://www.nal.usda.gov/wqic/Bibliographies/hypera.html.

———. "National Invasive Species Information Center." Available at http://www.invasivespeciesinfo.gov.

U.S. Department of the Army. *U.S. Army Survival Manual FM 21–76*, 1998. Distributed by Dorset Press, New York.

U.S. Department of Energy, Energy Efficiency and Renewable Energy. "Fueleconomy.gov." Available at http://www.fueleconomy.gov.

U.S. Forest Service. "Blue Oak." Available at http://www.na.fs.fed.us/pubs/silvics_manual/volume_2/quercus/douglasii.htm.

U.S. Forest Service, Rocky Mountain Research Station. "Restoration Ecology of Disturbed Lands." Available at http://www.fs.fed.us/rm/logan/4301.

U. S. Green Building Council. "Buildings and Climate Change." Available at http://www.usgbc.org/DisplayPage.aspx?CMSPageID=1617.

Van der Leeuw, Sander E., François Favory, and Jean-Jacques Girardot. "The Archaeological Study of Environmental Degradation: An Example from Southwestern France." Chapter 5 in Charles L. Redman, Steven R. James, Paul R. Fish, and J. Daniel Rogers, eds., *The Archaeology of Global Change: The Impact of Humans on Their Environment*. Washington, D.C.: Smithsonian Institution Press, 2004.

Vaughan, David G., and Robert Arthern. "Why Is It Hard to Predict the Future of Ice Sheets?" *Science* 315 (2007): 1503–1504.

Velicogna, Isabella, and John Wahr. "Measurements of Time-Variable Gravity Show Mass Loss in Antarctica." *Science* 311 (2006): 1754–1756.

Visser, Marcel E., and Leonard J. M. Holleman. "Warmer Springs Disrupt the Synchrony of Oak and Winter Moth Phenology." *Proceedings of the Royal Society of London B (Biology)* 286 (2001): 289–294.

————, and Phillip Gienapp. "Shifts in Caterpillar Biomass Phenology due to Climate Change and Its Impact on the Breeding Biology of an Insectivorous Bird." *Oecologia* 147 (2006): 164–172.

Wagner, Thomas. "Business World Finally Sees Potential Profits in Joining the Battle against Global Warming." Environmental News Network, August 17, 2007. Available at http://www.enn.com.

Wald, Matthew L., and Alexei Barrionuevo. "Chasing a Dream Made of Weeds: As Corn Prices Soar, Pressure Builds to Make a Cheap Cellulosic Ethanol." *New York Times*, April 17, 2007.

Walker, G. K., Y. C. Sud, and R. Atlas. "Impact of the Ongoing Amazonian Deforestation on Local Precipitation: A GCM Simulation Study." *Bulletin of the American Meteorological Society* 76 (1995): 346–361.

Wallis, Daniel. "Global Warming to Hit Poor the Worst, Says U.N.'s Ban." Environmental News Network, February 6, 2007. Available at http://www.enn.com.

Ward, Peter D. *Under a Green Sky: Global Warming, the Mass Extinctions of the Past, and What They Can Tell Us about Our Future*. New York: HarperCollins, 2007.

Ward, Peter D., et al. "Abrupt and Gradual Extinction among Late Permian Land Vertebrates in the Karoo Basin, South Africa." *Science* 307 (2005): 709–714.

Ward, Peter D., and Alexis Rockman. *Future Evolution: An Illuminated History of Life to Come*. New York: Henry Holt, 2001.

Wardle, David A., et al. "The Response of Plant Diversity to Ecosystem Retrogression: Evidence from Contrasting Long-Term Chronosequences." Oikos 117 (2008): 93–103.

Weatherford, Jack. *Indian Givers: How the Indians of the Americas Transformed the World*. New York: Ballantine, 1989.

Webster, P. J., et al. "Changes in Tropical Cyclone Number, Duration, and Intensity in a Warming Environment." *Science* 309 (2005): 1844–1846. Summarized by Richard A. Kerr, "Is Katrina a Harbinger of Still More Powerful Hurricanes?" *Science* 309 (2005): 1807.

Weiss, Ehud, Mordechai E. Kislev, and Anat Hartmann. "Autonomous Cultivation before Domestication." *Science* 312 (2006): 1608–1610.

West, Neil E. "Intermountain Deserts, Shrub Steppes, and Woodlands." Chapter 7 in Michael G. Barbour and William Dwight Billings, eds., *North American Terrestrial Vegetation*. New York: Cambridge University Press, 1988.

Westerling, A. L., et al. "Warming and Earlier Spring Increase Western U.S. Forest Wildfire Activity." *Science* 313 (2006): 940–943. Summary by Steven W. Running, "Is Global Warming Causing More, Larger Wildfires?" *Science* 313 (2006): 927–928.

Wieland, Nancy K., and Fakhri A. Bazzaz. "Physiological Ecology of Three Codominant Successional Annuals." *Ecology* 56 (1975): 681–688.

Wigley, T.M.L. "A Combined Mitigation/Geoengineering Approach to Climate Stabilization." *Science* 314 (2006): 452–454.

William J. Clinton Foundation. "Clinton Climate Initiative." Available at http://www.clintonfoundation.org/cf-pgm-cci-home.htm.

Williams, C. M. "Nutritional Quality of Organic Food: Shades of Grey or Shades of Green?" *Proceedings of the Nutrition Society* 61 (2002): 19–24.

Bibliography

Williams, Michael. *Deforesting the Earth: From Prehistory to Global Crisis.* Chicago: University of Chicago Press, 2006.

Williamson, Elizabeth. "Interior Department Official Facing Scrutiny Resigns." *Washington Post*, May 2, 2007.

Willis, K. J., L. Gillson, and T. M. Brncic. "How 'Virgin' Is Virgin Rainforest?" *Science* 304 (2004): 402–403.

Wilson, Edward O. *The Diversity of Life.* Cambridge, Mass.: Harvard University Press, 1992.

———. *The Future of Life.* New York: Knopf, 2002.

Wirzba, Norman, ed. *The Essential Agrarian Reader: The Future of Culture, Community, and the Land.* Lexington: University of Kentucky Press, 2003.

World Agroforestry Centre. "Agroforestry Addresses Climate Change and Poverty." Available at http://www.worldagroforestry.org.

Wynn, Gerard. "World Bank Eyes $250 Million Deal to Save Forests." Environmental News Network, May 4, 2007. Available at http://www.enn.com.

Yoon, Carol Kaesuk. "As Mt. St. Helens Recovers, Old Wisdom Crumbles." *New York Times*, May 16, 2000.

Zabarenko, Deborah. "U.N. Climate Change Meeting Aims at Rich Countries." Environmental News Network, July 31, 2007. Available at http://www.enn.com.

Zeng, N. "Drought in the Sahel." *Science* 302 (2003): 999–1000.

Zhang, Xiaoyang, et al. "Diverse Responses of Vegetation Phenology to a Warming Climate." *Geophysical Research Letters* 34 (2007). Available at http://www.agu.org/pubs/crossref/2007/2007GL031447.shtml.

Zimmerman, Martin H. *Xylem Structure and the Ascent of Sap.* Berlin: Springer-Verlag, 1983.

index

about the author

Stanley A. Rice is associate professor of biology at Southeastern Oklahoma State University in Durant. A native of Oklahoma, he grew up in California, obtained a B.A. in environmental biology from the University of California, Santa Barbara, and a Ph.D. in plant biology from the University of Illinois at Urbana-Champaign. He taught biology at colleges in New York, Indiana, and Minnesota before returning to Oklahoma in 1998, when he joined the faculty of Southeastern. His *Encyclopedia of Evolution* was published in hardcover in 2006 and paperback in 2007. He has published scholarly papers and given numerous presentations at professional meetings. Information about his research and writings can be found at http://www.stanleyrice.com.